PRINCIPLES AND PRACTICE

Shirley Pease
Carlos Lois (Eds.)

Mammalian and Avian Transgenesis – New Approaches

With 49 Figures, 12 in Color

Springer

Shirley Pease
California Institute of Technology
Division of Biology, 147-75
Pasadena
California 91125
USA

e-mail: ssp@its.caltech.edu

Carlos Lois
Massachusetts Institute of Technology, E25-421
77 Massachusetts Avenue
Cambridge, Massachusetts 02139
USA

e-mail: los@mit.edu

Library of Congress Control Number: 2005930817

ISBN-10 3-540-28415-X Springer Berlin Heidelberg New York
ISBN-13 978-3-540-28415-4 Springer Berlin Heidelberg New York

Springer is a part of Springer Science+Business Media
springeronline.com

Printed in Germany

Cover design: *design & production*, Heidelberg, Germany
Typesetting and production: LE-TEX Jelonek, Schmidt & Vöckler GbR, Leipzig, Germany
31/3150 YL 5 4 3 2 1 0 - Printed on acid-free paper

Preface

Now that the genomes of several animal species have been sequenced, the need to manipulate the genome in order to understand gene function has stimulated an even wider field of research. The impact of the techniques of gene manipulation on the understanding of developmental biology and disease in humans, as well as in other species, is exciting.

Since the first transgenic animal was made by Palmiter and Brinster in 1982 and the first targeted mutation by Capecchi, Evans and Smithies in 1989, the advancement of some of the technical aspects of transgenic technology has facilitated more precise manipulation of the mammalian genome and control of transgene expression. After these two initial developments came the application of the same, albeit modified, techniques to other species, such as cows, pigs and sheep. There then followed cloning and the understanding that differentiated cells, such as cumulus mass cells, could be re-programmed and returned to the embryonic environment in a pluripotent state. Via this enhancement of existing technology, we now see gene expression controlled in a more precise way, with technology that facilitates switching promoter genes on and off at precise times. Also, as methods allowing the delivery of transgenes become more varied, it has become possible to apply transgenic technology to more and more species, allowing the investigator to select the animal model most suited to the study of physiological processes in his or her field of interest. For example, the rat is most suited for cardiovascular research, the bird for early embryonic work.

It is clear that, in the future, cloning (nuclear transfer) will be very important in the advancement of the understanding of gene function. This is the first book in which current techniques of nuclear transfer are presented in a practical manner. This is also the first book to describe the procedures required to generate transgenic birds. The focus of this manual is to present information in a "user friendly" manner. Most chapters include detailed protocols on their subject matter. We anticipate that this manual will be a useful addition to any working laboratory employing such procedures.

Contributors to this manual are all actively working, in the course of their own research, in the development of these techniques. In alphabetical order, they are:

Niels Adams is a Senior Staff Scientist at Regeneron Pharmaceuticals where he runs the Gene Expression Laboratory, responsible for expression analysis of mice produced by the VelociGene program. He received his DPhil in Anatomy from Oxford University in 1995 and continued his research into developmental neurobiology at King's College, London, and Rockefeller University. Whilst at Rockefeller Niels was also involved in the GENSAT project, a high throughput gene expression screen of genes expressed in the CNS.

Kevin Eggan received his Ph.D. in Biology from the Massachusetts Institute of Technology in February of 2003. He is currently an Assistant Professor of Molecular and Cellular Biology at Harvard University. Dr. Eggan has devoted the last 7 years to performing stem cell research. He is currently leading a research group investigating the mechanisms regulating epigenetic reprogramming after somatic cell nuclear transfer, and using nuclear transfer techniques to derive disease-specific human embryonic stem cell lines from diabetic and Parkinson's patients. His accomplishments include cloning mice from olfactory sensory neurons, deriving embryonic germ cells and male gametes from embryonic stem cells, and characterizing the abnormalities that sometimes arise as a result of nuclear transfer.

Nicholas Gale received his PhD in Cellular and Developmental Biology from Stony Brook University in a joint Ph.D. program with Cold Spring Harbor Laboratory. He began his scientific career at Regeneron Pharmaceuticals, Inc. in New York in 1994 as a post doc with Dr. George D. Yancopoulos, and has remained with Regeneron, where he is currently Senior Director of Functional Genomics. Dr. Gale's main research interests are in understanding normal and pathological development of the blood and lymph vascular systems, and his research has relied heavily on the use of engineered mouse genetic models. He is a co-developer of Regeneron's VelociGene technology, which allows rapid and precise development of such models. In addition to leading a research group in vascular biology, he established and oversees Regeneron's Gene Expression Group, which characterizes reporter gene expression patterns and performs first line phenotypic analysis on new mouse models.

Schiaoching Gong received her Ph.D. from SUNY Health Science Center at Brooklyn in 1990. She is currently a Research Associate Professor at Rockefeller University. Dr. Gong has devoted the last 12 years to the generation of transgenic and knockout mice. She is the Director of the Molecular Biology Core in the GENSAT Project. Her accomplishments include the generation of transgenic mice used to map gene expression in the brain, and the development of a highly efficient system for the construction of targeting vectors from BACs for conditional knock-out and knock-in mutations and the application of this technology to the generation of mouse models for several human genetic disorders.

Mansuo Hayashi is a postdoctoral fellow in Dr. Susumu Tonegawa's laboratory at the Picower Institute for Learning and Memory at MIT. Dr. Hayashi received her Ph.D. in Molecular Biology at Princeton University. Her current work uses mouse genetic approaches to investigate molecular and cellular mechanisms underlying neurological diseases, including memory dysfunction, mental retardation and autism.

Shigemi Hayashi is a Research Investigator in the Models of Disease Center at the Novartis Institute for Biomedical Research. Prior to this current position, he was an associate research scientist in the Department of Neurology at Colombia University and a postdoctoral associate in the Department of Molecular and Cellular Biology at Harvard University. He received an M.S. in Pharmacy at Tokyo College of Pharmacy and a Ph.D. in Molecular Biology at Princeton University. His major interest resides in understanding molecular mechanisms of developmental biology, physiology and pathology in the mammalian nervous system using the mouse as a model organism.

Rudolf Jaenisch received his MD in 1968 at the University of Munich and became a postdoctoral fellow at Princeton in the laboratory of Arnold Levine, working on SV40 DNA replication and transformation. During this time he was also a visiting scientist in the laboratory of Beatrice Mintz, where he generated mice carrying SV40 sequences in all tissues. In 1972 he established his own laboratory at the Salk Institute in California but moved in 1977 to the Heinrich Pette Institute in Hamburg as head of the group on tumor virology. In 1984 he came to Boston as a founding member of the Whitehead Institute, and is currently Professor of Biology at Massachusetts Institute of Technology.

Dr. Jaenisch generated the first transgenic mice carrying exogenous DNA in the germ line and was the first to use insertional mutagenesis to identify genes crucial for embryonic development. Perhaps his most fundamental contributions have been in the study of epigenetic processes during development. In particular, he showed that methylation of DNA plays important roles in gene expression, imprinting and X-inactivation, as well as in diseases such as cancer and mental retardation. Dr. Jaenisch's pioneering work in the generation of transgenic mice has produced important advances in the understanding of cancer, neurological and connective tissue diseases, and developmental abnormalities. Mouse models generated in the laboratory of Dr. Jaenisch have been used to explore basic questions such as the role of DNA modification, genomic imprinting, X chromosome inactivation, and, most recently, the nature of stem cells, epigenetic reprogramming and nuclear cloning. These studies have had a major impact on the current debate regarding cloning and its use in the potential treatment of disease. His laboratory is currently gaining insights into therapeutic cloning, and has indeed rescued mice having a genetic defect through therapeutic cloning and gene therapy.

Rebecca Krisher received her Bachelor's degree in Biology from Hanover College followed by a Master's degree in Animal Sciences at North Carolina State University. She then worked for Granada Biosciences before completing her Ph.D. at Virginia Tech in Animal Science/Dairy. Dr. Krisher then worked as an embryologist in the field of human infertility before returning to academia as a post-doc at the University of Wisconsin-Madison. Dr. Krisher joined the faculty of the Department of Animal Sciences at Purdue University in 1998, where she is currently an Associate Professor. With her background in both academic and commercial aspects of assisted reproductive technology, Dr. Krisher's lab maintains a strong interest in domestic animal biotechnology. Dr. Krisher's research program focuses on defining the complex processes that occur within female gametes as they develop and mature before ovulation, as well as before and after fertilization. Changes that occur during maturation of the oocyte include both meiotic progression and acquisition of developmental competence. Cellular events occurring during oocyte growth and maturation have an enormous influence on the pre- and post-implantation development of the resulting zygote. Dr. Krisher's laboratory is elucidating the changes in oocyte cytoplasm during growth and maturation that are critical components of developmental competence after fertilization.

Ming-Wen Li is a specialist in ICSI microinjection and other services in the Murine Targeted Genomics Laboratory at UC Davis. Dr. Li received his Ph.D. degree in physiology at the Chinese Academy of Sciences in Beijing. He has 18 years experience as a scientific researcher in the field of fertilization biology.

K. C. Kent Lloyd is director of the Murine Targeted Genomics Laboratory, Associate Professor at the Center for Comparative Medicine, and Associate Dean of Research and Graduate Education in the School of Veterinary Medicine, UC Davis. Dr. Lloyd received his DVM from UC Davis and his Ph.D. in physiology at UCLA. He has more than 20 years experience in physiology research and teaching and specializes in the development and application of animal models of human and animal diseases.

Carlos Lois received his MD from the University of Valencia (Spain) and his Ph.D. from the Rockefeller University. He is currently an assistant professor in Neurobiology in the department of Brain and Cognitive Sciences at the Massachusetts Institute of Technology. His laboratory studies the mechanisms of neuronal addition into the brain of adult animals by using gene-transfer techniques.

Shirley Pease is currently a Member of Professional Staff at the California Institute of Technology in Pasadena, California. Originally from England, she started her career in laboratory animal science in the early 1970s and since then has been employed in several European institutions of note, such as the Laboratory Animals Centre, Carshalton, Surrey, UK; St Mary's Hospital Medical School, Paddington, London; the Institute of Cancer Research, Belmont, Surrey, UK and the European Molecular Biology Laboratory, Heidelberg, Germany. She was part of a team that established the use of leukaemia inhibitory factor in the culture of embryonic stem cells in the late 1980s and, more recently, in developing the system for gene delivery by way of lentiviral vectors.

Robert M. 'Bob' Petters is Professor of Animal Science at North Carolina State University. Bob completed his Ph.D. in Genetics at NC State in 1976. During his time as a postdoc at Yale University, he was introduced to the biology of the mammalian preimplantation embryo and completed experiments using chimeric mice. Returning to NC State in 1983, Bob began applying embryo manipulation technologies in his work with pig embryos, first to the in vitro culture of embryos and later to DNA microinjection. The development of the culture medium NCSU 23, which continues to be used by many labs, was a major accomplishment of Petters' lab. More recently, Bob's lab has made a genetic model of the human disease retinitis pigmentosa (RP), using transgenic pigs. RP is an inherited form of blindness for which there is no treatment. Research continues on genetic models of human disease using transgenic pigs. Bob lives in Cary, NC, with his wife of 32 years, Marcia.

William Ritchie gained a qualification in Agriculture at Edinburgh University in the early 1970s and then went to work on one of the then "Animal Breeding Research Organisations" (now "The Roslin Institute") farms in the Scottish Borders, recording data from animal experiments. At this time he began work on an Honours Degree in "The Open University". Bill started his training in Nuclear Transfer in 1989 and produced his first cloned sheep in 1992. Morag and Megan, the first animals produced from cultured cells, were produced at Roslin in 1995. This allowed the group to use this technique in the production of seven cloned animals from cultured cells in 1996, one of which was "Dolly", the first animal cloned from an adult cell. They further developed the technique, producing the first transgenic animals by nuclear transfer (NT), in 1997. These animals contained a gene for "human blood factor IX", which was expressed in the milk of these sheep. Bill was member of the team that produced NT lambs containing a deletion of the PrP gene in 2001, and the production of the first cloned pig in Europe in 2002. More recent work involves the use of lentivirus in the making of transgenic pigs.

Benjamin Scott is a graduate student in the department of Brain and Cognitive Sciences at the Massachussetts Institute of Technology. His research focuses on the development of methods for genetic manipulation of birds.

Pasadena, August 2005 *Shirley Pease*

References

Palmiter RD, Brinster RL, Hammer RE, Trumbauer ME, Rosenfeld MG, Birnberg NC, Evans RM (1982) Dramatic growth of mice that develop from eggs microinjected with metallothionein-growth hormone fusion genes. Nature 300: 611-615

Capecchi MR (1989) Altering the genome by homologous recombination. Science 244 (4910):1288-92

Doetschman T, Gregg RG, Maeda N, Hooper ML, Melton DW, Thompson S and Smithies O (1987) Targetted correction of a mutant HPRT gene in mouse embryonic stem cells. Nature 330:576-578

Contents

Contributors

Adams, Niels C.
Regeneron Pharmaceuticals, 777 Old Saw Mill River Road, Tarrytown, NY 10591, USA

Eggan, Kevin
The Stowers Medical Institute, Department of Molecular and Cellular Biology, Harvard University, Cambridge, MA 02138, USA, e-mail: Eggan@mcb.harvard.edu

Gale, Nicholas W.
Regeneron Pharmaceuticals, 777 Old Saw Mill River Road, Tarrytown, NY 10591, USA, e-mail: nicholas.gale@regeneron.com

Gong, Shiaoching
Laboratory of Molecular Biology, Howard Hughes Medical Institute, Rockefeller University, Box 260, 1230, NY 10021, USA, e-mail: gongs@mail.rockefeller.edu

Hayashi, Mansuo L.
Department of Biology and the Picower Institute of Learning and Memory, Massachusetts Institute of Technology, Cambridge, MA 02139, USA, e-mail: mansuo@mit.edu

Hayashi, Shigemi
Models of Disease Center, Novartis Institutes for BioMedical Research, 250 Massachusetts Ave, Cambridge, MA 02139, USA, e-mail: shigemi.hayashi@novartis.com

Heintz, Nat
Laboratory of Molecular Biology, Howard Hughes Medical Institute, Rockefeller University, Box 260 1230 New York 10021, USA e-mail: heintz@rockvax.rockefeller.edu

Jaenisch, Rudolf
Whitehead Institute for Biomedical Research and Department of Biology, Massachusetts Institute of Technology, 9 Cambridge Center, Cambridge, MA 02142, USA

Krisher, Rebecca L.
Department of Animal Sciences, Lilly Hall of Life Sciences, Lab VPTH 201/204, 915 West State St., Purdue University, West Lafayette, IN 47907–2054, USA

Li, Ming-Wen
Center for Comparative Medicine, School of Veterinary Medicine, University of California, One Shields Avenue, Davis, CA 95616, USA

Lloyd, K.C. Kent
Center for Comparative Medicine, School of Veterinary Medicine, University of California, One Shields Avenue, Davis, CA 95616, USA, e-mail: kclloyd@ucdavis.edu

Lois, Carlos
Massachusetts Institute of Technology, E25–421, 77 Massachusetts Avenue, Cambridge, MA 02139, USA, e-mail: los@mit.edu

Pease, Shirley
California Institute of Technology, Division of Biology, Pasadena, CA 91125, USA, e-mail: ssp@its.caltech.edu

Petters, Robert M.
Department of Animal Science, North Carolina State University, Box 7621, Raleigh, NC 27695–7621, USA, e-mail: bob_petters@ncsu.edu

Ritchie, William A.
Department of Gene Expression and Development, Roslin BioCentre, Midlothian EH25 9PS, Scotland, UK, e-mail: william.ritchie@bbsrc.ac.uk

Scott, Benjamin B.
Massachusetts Institute of Technology, E25–421, 77 Massachusetts Avenue, Cambridge, MA 02139, USA, e-mail: bbscott@mit.edu

1 Generation of Transgenic Animals Using Lentiviral Vectors

Carlos Lois

1.1 Introduction

The production of transgenic animals has been one of the most powerful technical advances in the understanding of genetics (Jaenisch 1988). Transgenic animals have been instrumental in improving our understanding of human genetic diseases and will probably be of increasing importance in the improvement of livestock, xenotransplantation, and the production of biologically active pharmaceuticals.

The first transgenic animals were generated by infecting mouse blastocyst cells with a Moloney leukemia virus-based retroviral vector (Jaenisch 1976). These animals were able to transmit the integrated proviral transgene to their progeny. However, neither the founders nor the subsequent offspring expressed the exogenously introduced gene at sufficient levels to be of practical use for most experimental applications. Subsequent studies showed that silencing of the proviral transgene was correlated to developmentally regulated methylation of the flanking host sequences (Jahner et al. 1982). To circumvent the developmental silencing observed with retroviral transgenics, a nonviral approach to transgenesis called pronuclear injection was developed (Gordon et al. 1980). In pronuclear injection, a linearized DNA sequence is introduced directly into the pronucleus of single-cell embryos by microinjection. Generally, transgenic animals that are generated by pronuclear injection transmit and reliably express the exogenous sequences, and this has become the standard approach for generating transgenic mice. Most commonly, the transgene integrates in head-to-tail tandem arrays of up to 100 copies at a single chromosomal site, and multiple copies of the transgene are frequently required to drive expression of the exogenous sequence to detectable levels (Brinster et al. 1985). Although pronuclear injection is widely used to generate transgenic mice, this procedure suffers from several limitations. First, because the technique requires visualization of the male pronucleus, which is easily seen only in the zygotes of certain strains of mice, pronuclear injection is

Carlos Lois: Massachusetts Institute of Technology, E25-421, 77 Massachusetts Avenue, Cambridge, MA 02139, USA, e-mail: los@mit.edu

Principles and Practice
Mammalian and Avian Transgenesis – New Approaches
Shirley Pease and Carlos Lois (Eds.)

not practical for transgenesis in other animals. Second, since pronuclear injection requires penetration of both the cell membrane and the pronuclear envelope for DNA delivery, it is technically challenging, requires sophisticated micromanipulation equipment, and is often damaging to the embryo. Finally, under optimal conditions, superovulated mammals yield 10–20 embryos/female and only 1% of microinjected mouse embryos, and 0.1% of cattle, sheep and pig embryos, develop into transgenic animals. Taken together, these limitations require the injection of hundreds of embryos to ensure the derivation of one transgenic line. While obtaining large numbers of embryos is possible in mice, this requirement constitutes a significant cost barrier to generating many different transgenic lines. Furthermore, these constraints seriously limit the utility of pronuclear injection in larger mammals that would, in some cases, be a more appropriate model for human diseases and more suitable for studying organ systems or behavior. In addition, the generation of transgenic birds has long been a goal, given the potential for producing large amounts of protein in their eggs. Unfortunately, pronuclear injection is not applicable to the generation of transgenic birds because the pronucleus is not visible in the avian embryo due to its high yolk content (Ivarie 2003).

We recently observed that lentiviral vectors can be used to generate transgenic animals, and that, in contrast to other retroviruses, lentiviral vectors are not developmentally silenced. The technique of transgenesis by lentiviral gene delivery to one-cell embryos overcomes many of the limitations of other techniques, and also provides for reproducible tissue-specific expression from promoters with restricted specificity. It is more efficient than pronuclear injection, with approximately 80% of surviving embryos transmitting the transgene to their progeny, and over 90% of transgenic founders expressing the transgene. In addition, the method is less invasive to the embryos, more cost-effective, and technically less demanding. Delivering lentiviruses by co-incubation with denuded embryos (as described below) obviates the need for micromanipulation and may be an easier option for many laboratories wanting to make transgenic animals. Furthermore, since the lentiviral delivery technique does not require visualization of the pronucleus, it has the potential to be extended to diverse mouse strains, as well as other animal species. For instance, using this technique we have generated transgenic rats, a species into which the introduction of exogenous genes has so far been difficult and inefficient.

The technique works equally well with all mammals because the vesicular stomatitis virus G (VSVg) protein that mediates viral entry finds receptors on the cells of most species, including cattle and primates (Hofmann et al. 2003, 2004). Furthermore, it allows for the germline introduction of exogenous genes into birds, a class of animals for which no satisfactory method for creating transgenics exists. The combination of a high efficiency of transgenesis, low cost, and scalability for high throughput, allows for other applications of the lentiviral transgenesis technique, such as fine-scale mutagenesis of regulatory sequences, large-scale insertional mutagenesis screens, or gene trapping. However, as with other retroviral vectors, use of lentiviruses for the purposes of transgenesis may be limited by processes that interfere with viral production, such as splicing or polyadenylation signals present in the transgene, or by the insertion of transgenes larger than 10 kB between the long terminal repeats (LTRs).

1.2 The Production of Lentiviral Vectors

1.2.1 Design and Construction of the Lentiviral Transfer Vector

(For a general reference on retroviral design, see (Lois et al. 2001). The basic design of the transfer vector consists of a cassette in which a promoter (P) drives the expression of a gene of interest (G) and this P+G cassette is inserted into a lentiviral vector flanked by LTR sequences (Lois et al. 2002). In most cases the engineering of a lentiviral transgene construct is a straightforward procedure. In some situations, however, some P+G combinations might be problematic and require some modifications. Several considerations have to be taken into account before engineering a lentiviral transfer vector. First, the capsid of the retrovirus has a maximum capacity of approximately 12 kB. Because the transfer vector contains LTRs and some sequences necessary for reverse transcription and packaging, the theoretical maximum size of the P+G cassette is 10 kB. Second, lentiviruses have an RNA genome and, therefore, the production of recombinant viruses might be affected by RNA processing mechanisms such as splicing and polyadenylation. Polyadenylation signals, such as those from Simian virus 40 (SV40), bovine growth hormone (BGH) or herpes simplex virus thymidine kinase (HSVtk), would interfere with production of the full-length RNA lentiviral genome, and should be removed from the inserts. (The 3′-LTR of the virus vector provides the polyadenylation signal for the integrated transgene). The presence of introns in the inserts might result in aberrant splicing of the lentiviral genome, with subsequent loss of fragments of the insert from the integrated transgene. Whenever possible, it is advisable to use cDNAs because the splicing signals have already been eliminated. However, the presence of introns in the promoter is not necessarily a problem. We have generated transgenic animals with promoters such as the human polyubiquitin-C promoter (UbiC), or cytomegalovirus (CMV)/β-actin (see below) that contain introns, and expression from these transgenes is appropriate. If the viral titer is low, or expression from the lentivirus is inadequate, and all other steps in the production of the virus (transfection, concentration) are correct, it is worthwhile to check whether mRNA processing has interfered with viral production. It is important to realize that splicing and polyadenylation signals are in some cases easily identified, but that cryptic signals might also be present. If introns or polyA addition signals are known to exist in the promoter and it is not possible to remove them, it might be necessary to generate a transfer vector in which the P+G cassette is inserted in the opposite orientation with respect to transcription of the lentiviral genome (5′-LTR to 3′-LTR). Note that some polyA addition signals, such as that of SV40, are bidirectional and could interfere with virus production even when inserted in the reverse orientation.

1.2.2 Promoters

For ubiquitous expression, the human UbiC promoter directs high and reliable expression in most tissues studied in both mice and rats (Lois et al. 2002). Other promoters, such as human phosphoglycerate kinase, chicken β-actin, the CMV enhancer/promoter or CMV enhancer/chicken β-actin promoter, do not provide reliable expression across all cell types. In addition, we have successfully generated transgenic animals with the myogenin promoter (specific for skeletal muscle) and the lck proximal promoter (specific for T-lymphocytes; (Lois et al. 2002)). It is advisable to use compact promoters (under 5 kB) because this will reduce the chances of splicing or polyadenylation signals being present in the promoter sequence.

1.2.3 Reporters

It is very convenient to include a reporter gene in the construct in order to quickly assess virus titer during production. Similarly, once the transgenic animals have been generated, the presence or absence of the reporter can be used to screen for positive transgenic animals. If no reporter gene is present in the construct, determining viral titer should be performed by immunocytochemistry, with antibodies raised against the overexpressed gene of interest.

We routinely include an internal ribosome entry site (IRES)-reporter gene cassette in our lentivirus constructs so that we can track expression of the gene of interest simply by looking for reporter gene expression (Lois et al. 2001; Martinez-Salas 1999). Currently, three reporter genes are widely used: green fluorescent protein (GFP), β-galactosidase (lacZ), and alkaline phosphatase (AP).

1.2.3.1 Green Fluorescent Protein

GFP offers several advantages as a reporter gene. First, it can be visualized in live cells both in vivo (by two-photon laser confocal microscopy) and in culture. Second, it can be identified directly by its intrinsic fluorescence, or its signal can be amplified by immunostaining with highly specific monoclonal antibodies. GFP is a small molecule (approx. 27 Kd) that, in its native state, distributes evenly throughout the cytoplasm. Furthermore, by adding specific amino acid sequences, variants have been constructed that localize GFP in several subcellular compartments (nucleus, mitochondria, membrane, etc.). The main disadvantage of GFP is that when cells are fixed with aldehyde fixatives, its intrinsic fluorescence fades significantly in the first 24–48 h. In these situations it is advisable to detect GFP-expressing cells by immunostaining with antibodies raised against GFP. Finally, it is important to note that GFP must reach a significant level of expression in order to be detected, and that cells that express low levels of GFP may be missed when examined for an intrinsic GFP signal. Again, staining with antibodies raised against GFP will often reveal cells that express GFP below the level of intrinsic fluorescence detection.

1.2.3.2 β-Galactosidase

The main advantage of lacZ as a reporter gene is its high sensitivity. Cells expressing lacZ can be incubated for days in the presence of lacZ substrates, thus producing a very high amplification of the signal. It is important to remember that some tissues (condrocytes, choroid plexus, macrophages, etc.) have some level of endogenous galactosidase activity that can be mistaken for lacZ staining. LacZ background staining can be, in most cases, eliminated or reduced with appropriate fixation (Shimohama et al. 1989).

1.2.3.3 Alkaline Phosphatase

AP, like lacZ, is also extremely sensitive and can be amplified for many hours. Some tissues exhibit some endogenous AP background activity, which can be reduced with the chemical levamisol (Fields-Berry et al. 1992).

1.2.4 Plasmid Preparation

The most common problem faced by researchers attempting to produce lentiviruses is recombination of the lentiviral vectors when growing the plasmids. DNA from lentiviral backbones occasionally does not grow well in some bacterial strains. It is advisable to transform all constructs into Escherichia coli strains 'Sure' or 'Topp10'. These strains have been engineered allow propagation of plasmids prone to recombination. Also, in order to prevent problems associated with abnormal plasmid growth, it is important to start cultures from a fresh bacterial streak (plasmid yield may be low when grown from cells kept at 4 °C on a plate). In addition, it is advisable to first grow a small culture (1 ml) of bacteria for 10 h before transferring the bacteria to a large-scale culture (200 or 500 ml). Finally, to ensure optimal transfection, it is important that the plasmid DNA is very clean: use Qiagen maxiprep columns or similar. To conveniently store the plasmid DNA, dilute it to 0.5 μg/μl in water, and freeze aliquots of approximately 200 μl at –20 °C.

1.2.5 Preparation of 293T Cells

The best cells for transfection are 293T cells, which can be obtained from ATCC, catalog number CRL 11268 (http://www.atcc.org/SearchCatalogs/longview.cfm?view=ce,646415,CRL-11268&text=293t&max=20).

1.2.5.1 Cell Thawing

Thaw the cells as follows: take a frozen vial from a liquid nitrogen container and immediately place the vial in a 37 °C water bath. The cells should thaw in less than 2 min. Without delay, transfer the cells to a 15 ml tube and, very gently, add dropwise 10 ml pre-warmed Dulbecco's modified Eagle's medium (DMEM) containing 10% fetal bovine serum (FBS) medium. It is important to maintain separation between the layer of the freezing medium that contains the cells and the serum-containing medium being added: the freezing medium contains 10–20% glycerol, and cells will lyse by osmotic shock if they are mixed suddenly with medium that does not contain glycerol. After layering the DMEM +10% FBS medium on top of the cells, spin for 5 min at 1,500 rpm in a table-top centrifuge. With a Pasteur pipette attached to a vacuum line, remove as much of the medium as possible without touching the cell pellet, which should be clearly visible at the bottom of the tube. Add 4 ml DMEM +10% FBS medium to the pellet and resuspend the cells by gently pipetting up and down 5–6 times. If the cells are pipetted too harshly, they will lyse. If the pipetting is not sufficient, cells will remain aggregated in large clumps and they will subsequently grow very slowly. When cells are appropriately resuspended, seed approximately 1–2 million cells on a 6 cm tissue culture plate and place the plate in a tissue culture incubator (humidified, 37 °C and 5% CO_2). Change the medium the next day and, if the cells are confluent, trypsinize and split into a 10 cm plate as detailed in the following section.

1.2.5.2
Cell Splitting

With a Pasteur pipette connected to a vacuum line, remove all the medium from the plate. Add 5 ml phosphate-buffered saline (PBS; room temperature) to the cells and swirl to spread over the whole surface of the plate. With a fresh Pasteur pipette, remove the PBS and add 1 ml cold trypsin (0.05% trypsin/EDTA; Gibco, Rockville, MD). Thawed trypsin should be kept at 4 °C at all times to prevent autodigestion and loss of activity in the enzyme. Do not warm the trypsin before applying it to cells, or the trypsin will degrade and the cells will not be appropriately dissociated for splitting. Swirl the trypsin solution to ensure complete spreading over the surface of the plate. Return the plate to the tissue culture incubator (at 37 °C) for 5 min. After 5 min the cells should be easily dislodged by tapping the side of the plate. If, after tapping, most cells remain attached to the plate, swirl again to cover the cells with trypsin and return the plate to the incubator for another 5 min. Tap again to dislodge the cells and add 10 ml prewarmed DMEM +10% FBS medium. Triturate aggressively by applying the tip of the pipette against the bottom of the tube: 293 cells are very hardy, and can withstand very harsh trituration.

1.2.5.3
Splitting for transfection

Because the production of virus is relatively time consuming, it is most practical to prepare a large stock of virus and keep it frozen in aliquots for future use. To prepare a stock of virus it is convenient to transfect cells in 15 cm plates. As a rule of thumb, one confluent plate should contain enough cells to seed three plates (of the same size) to be transfected the following day. Before splitting the cells, precoat the plates with a 0.2% gelatin solution. (To prepare the gelatin solution add 1 g gelatin to 500 ml water and autoclave. Store sterile, at room temperature). Gelatin-coated plates can be prepared in advance and stored at 4 °C or at room temperature for later use. For a 15 cm plate, add 100 ml gelatin solution, with a plastic serological pipette, and swirl the plate to spread evenly. With the same plastic pipette, aspirate the medium from the first plate and transfer the gelatin solution to the next plate. Repeat this procedure until all plates are coated, then remove the residual liquid with a Pasteur pipette connected to a vacuum and leave the plates open in the tissue culture hood for 20 min until they dry completely. Once the plates are completely dry, trypsinize the cells as indicated above, count the cells and dilute them to 0.5×10^6 cells/ml. This cell density is important as 293T cells do not grow well after transfection. If they are seeded at a lower density, they will not reach confluence and will die. Some plasmids are troublesome for retrovirus production. For instance, some cDNAs are quite toxic to 293T cells upon transfection. In such cases it is very important that the cells are close to confluent (at least 0.5×10^6/ml upon plating) just before transfection. One of the most common reasons for low virus titer is inadequate confluency of the cells upon transfection: too confluent and transfection will be inefficient, too sparse and the cells die.

1.2.5.4 Seeding of 293T cells

- 10 cm plates Dilute the cells into 12 ml (at 0.5×10^6 cells/ml) medium.
- 15 cm plates Dilute the cells into 25 ml (at 0.5×10^6 cells/ml) medium.

Seed the cells on gelatinized plates and swirl the plate randomly to ensure even spreading of the cells. It is important to prevent the clumping of the cells in the center of the plate. This will occur if the plates are moved in a circular motion; move plates in random directions to prevent regional clumping.

1.2.6 Transfection

At 16–20 h after seeding, remove 2 ml (from a 10 cm plate) or 5 ml (from a 15 cm plate) of medium. Leave the cells in the incubator for at least 1 h. Do not move the plates outside the incubator after medium removal (the pH will change and transfection will not be efficient).

1.2.6.1 Transfection reagents

The quality of the transfection reagents is probably the main determinant for obtaining high viral titers. Pay special attention to the preparation of the HBSS solution (see below). Because the sodium phosphate component is included in a small amount (less than a gram per liter), use an accurate balance capable of measuring milligrams when weighing it. Calibrate the pH meter before adjusting the pH of the solution. Leave the solution stirring for 10 min after the pH of 7.05 has been reached, to ensure that it does not drift. Once the pH is stable, sterilize by filtering the solution through a 0.2 μm filter. HBSS prepared in this way can be kept at room temperature for at least a year. We usually keep the 2 M calcium chloride solution frozen at −20 °C in 5 ml aliquots and thaw immediately before transfection. It is also possible to transfect 293T cells with commercial reagents, such a lipofectamine, lipofectin, etc.; however, these reagents are expensive. In addition, they may be toxic for the target cells infected with the concentrated virus.

HBSS	solution
0.28 M	NaCl
0.05 M	Hepes buffer
1.5 mM	Na_2HPO_4

- fill up to 500 ml deoinized water
- titer to exactly pH 7.03–7.04 with NaOH
- filter sterilize through a 0.22 μm filter

Preparation of the transfection mix. On the bench, mix the DNA, water and calcium chloride (sterile technique not required) in 15 ml polypropylene tubes.

Volume of mix per plate. For a 10 cm plate, mix the following in a 15 ml polypropylene tube: 20 μl transfer vector, 15 μl delta 8.9 plasmid (the packaging plasmid), plus 10 μl VSVg plasmid (the pseudotype envelope plasmid). Add 130 μl 2 M $CaCl_2$ and fill up to 1 ml with distilled-deionized $(dd)H_2O$.

For a 15 cm plate. Mix 40 μl transfer vector, 30 μl delta 8.9 plasmid, plus 20 μl VSVg plasmid. Add 260 μl 2 M $CaCl_2$ and fill up to 2 ml with ddH_2O.

1.2.6.2
Transfection procedure

In the hood, in separate 15 ml polypropylene tubes, prepare 1 ml (for a 10 cm plate), or 2 ml (for a 15 cm plate) 2× HBSS pH 7.05. It is important to perform this step in the hood in order to prevent contamination of the

HBSS stock. Immediately before transfecting, prepare all the tubes with the transfection components and arrange the pipette aids and Pasteur pipettes in the hood. Take one plate out of the incubator and place it in the hood. Mix the tube containing DNA+$CaCl_2$+H_2O well, pipette up and add it to the tube containing HBSS. Pipette the DNA+$CACl_2$+HBSS aggressively five times. Pipette the mixture onto the side of the plate while swirling and then put the plate back in the incubator. Repeat this for the other plates. It is important to minimize the time that each plate is outside of the tissue culture incubator in order to prevent significant pH changes, which will reduce transfection efficiency. At 3.5–5 h after adding DNA+$CaCl_2$+HBSS mix to the cells, aspirate the medium, wash twice with PBS and add 10 ml (10 cm plate) or 20 ml (15 ml plate) fresh DMEM+10% FBS medium. (If cells are not washed with PBS, some of the precipitate might remain attached to the plastic and many of the cells will die). The plates are then returned to the incubator and left there for 60–72 h, after which the supernatant is collected. If the transfer vector contains GFP or other fluorescent reporter, cells can be examined 24 h after transfection by live microscopy. This is also a convenient time to check the health status of the 293 cells under the microscope; 48 h after transfection, more than 90% of the cells should be transfected. If the transfection efficiency is low, re-check the quality of the DNA, the HBSS, or the 293T cells. The two most common reasons for poor transfection of 293T cells are: (1) the cells are contaminated by mycoplasma, or (2) the cells have been abused and allowed to become over-confluent during previous passages. If either of these situations has occurred, it is advisable to obtain a new seed of 293T cells from ATCC. Another reason for poor transfection efficiency might be the toxicity of some constructs when expressed at high levels in 293 cells, which will prevent the transfected cells 293T cells from becoming confluent.

1.2.7 Viral Concentration

At 60–72 h after transfection, the cells should be completely confluent and the supernatant ready for collection. Wash the ultracentrifugue tubes with 100% ethanol and invert them over a Kimwipe tissue, to drain off the ethanol. Wash the tubes with PBS or medium. To pre-cool the ultracentrifuge, turn it on, set the temperature to 4 °C and start applying a vacuum. It will take at least 30 min before the centrifuge reaches 4 °C.

To collect the viral supernatant, remove the plates from the incubator and pipette the medium into 50 ml conical tubes. Spin the tubes at 2,000 rpm in a table-top centrifuge for 5 min to remove large particles of cellular debris. In the hood, put 20 ml DMEM+10% FBS medium into a 0.45 μm self-contained filter (Nalgene; www.nalgenenunc.com). Apply a vacuum and discard the filtered medium. This prewetting of the filter reduces unspecific binding of viruses to the filter that would otherwise reduce the viral titer. Immediately after pre-wetting, pipette the virus suspension onto the filter and apply a vacuum until the suspension has passed through the filter. Pipette the filtered suspension into 40 ml ultraclear centrifuge tubes (Beckman, Fullerton, CA). In order to prevent spilling during spinning, seal the tubes with Parafilm. Place the sealed tubes in ultracentrifuge buckets. If necessary, put tubes filled with an equal volume of water in other buckets to balance the rotor. It is very important that tubes are correctly balanced due to the high speed at which the ultracentrifuge operates. Make sure that the difference in weight between the tubes is less than 0.1 g, and that buckets are tightly positioned in the rotor in the specified slots before installing the rotor in the centrifuge.

Program the centrifuge to spin for 90 min at 4 °C, 25,000 rpm. Watch the centrifuge until the rotation speed reaches at least 3,000 rpm. In the event that the centrifuge produces an error signal, stop immediately and confirm that the buckets are appropriately placed in their slots. Also weigh the tubes to ensure that they are correctly balanced. When the spin is finished, disconnect the vacuum of the ultracentrifuge, remove the buckets from the rotor and transfer them to the tissue culture hood. Prepare a 500 ml beaker containing 10 ml bleach and pour the viral supernatant into it. Invert the tubes over a sterile Kimwipe tissue to drain the remnants of the medium. It is usually very difficult to detect a pellet at the bottom of the tube. If a pellet is clearly visible, this probably indicates that an excess of cellular debris is present in the supernatant. Excessive cellular debris usually indicates that many of the cells died after transfection. With a Pasteur pipette, aspirate the supernatant exhaustively, until the tube seems dry. Add 100 μl cold PBS, (no bicarbonate, +Ca, +Mg), to the bottom of the tube. Seal the tubes with Parafilm to prevent evaporation. Keep the tubes at 4 °C for at least 12 h in order to dissolve the pellet. It is advisable to keep the tubes moving overnight in a rocker, or bellydancer, to resuspend the virus thoroughly. After 12–24 h at 4 °C, pipette the medium up and down gently to dissolve the pellet. Dispense the virus into 10 μl aliquots in Eppendorf tube for immediate use or to freeze at −70 °C. Write the date of harvest on the top of the tube. Two steps are important for achieving successful resuspension of the virus. Firstly, make sure that all of the DMEM+10% FBS medium is aspirated from the tube; FBS may contain components that can inactivate the virus when incubated at 4 °C. Secondly, the virus is relatively frail and is easily inactivated by aggressive pipetting. Thus it is particularly important to prevent any kind of frothing at this step.

1.2.8 Virus Titration

Twenty-four hours before titration, resuspend 293T cells at $0.5 \times 106/$ml (or HeLa cells at 5×10^5/ml) and add 100 μl cell suspension to each well of a 96-well plate. For viruses with tissue-specific promoters, use a cell type in which the particular promoter is active. On the day of infection, prewarm 10 ml DMEM +10% FBS. Prepare six Eppendorf tubes for titration. Add 90 μl DMEM +10% FBS to each tube. Add 10 μl virus solution to the first tube. Mix thoroughly, change the pipette tip and take 10 μl to mix with the next tube. Repeat the procedure until you reach the sixth tube, thus preparing suspensions of virus at sequential dilutions, or viral titer. Aspirate all the medium from the wells of the 96-well plate. Add 45 μl of each viral dilution to each of six wells (for analysis of duplicates), and incubate for at least 8 h before replacing the DMEM+10% FBS medium. Check for GFP fluorescence or for immunocytochemistry 3 days after infection.

Example: 20 colonies of GFP+ cells in the sixth dilution would be equal to 2×10^6 infectious units per 5 μl, or 4×10^5 infectious units per 1 μl [10 μl virus solution diluted 100,000 times produced 20 colonies (in duplicate wells), 5 μl would produce 2×10^6 colonies, thus 1 μl would contain 4×10^5 infectious viral particles]. A reasonable titer after concentration is $0.5 - 5 \times 10^6$ pfu/μl. If visible reporters such as GFP, lacZ or AP are not present in the virus, titrate by immunocytochemistry with antibodies raised against the gene driven by the internal promoter.

If there are no established cell lines in which the promoter is active, an approximate titration can be obtained by infecting primary cells, such as fibroblasts obtained from an animal, and maintaining them in culture for a few days. If there are no specific antibodies that can detect the expression of the transgene, a viral titer can be determined by ELISA against the matrix protein of the virus, or by quantifying reverse transcriptase activity in the viral concentrate.

1.3 Production of Transgenic Mice and Rats

The generation of transgenic animals with lentiviral vectors involves the use of some procedures, such as pronuclear injection and blastocyst transfer, that are common to other transgenic techniques. All aspects of superovulation, embryo collection and implantation are described in Chap. 11, *Ancillary Techniques*.

1.3.1 Superovulation and Embryo Collection

Female mice and rats are administered gonadotropins prior to mating to increase the number of eggs that are released from the ovaries at one time, a technique called superovulation. Female mice and rats are superovulated with a combination of pregnant mare's serum (PMS) and human chorionic gonadotropin (hCG). Prepubescent female mice around 25 days of age and weighing between 12.5 and 14 g are injected intraperitoneally (IP) with 5 IU PMS (Sigma G-4527, 25 IU/ml in 0.9% NaCl; Sigma, St. Louis, MO) between 1:00 and 3:00 p.m. on day –2, followed by 5 IU of hCG (Sigma C 8554, 25 IU/ml in 0.9% NaCl) 48 h later on day 0. Prepubescent female rats between 28 and 30 days of age and weighing between 70 and 80 g are injected IP with 25 IU PMS between 1:00 and 3:00 p.m. on day –2, followed by 5 IU of hCG 48 h later on day 0. For both rats and mice, hormone-treated females are caged with fertile males 2–3 months of age in mice and 3–4 months of age in rats, for mating overnight. On the morning of day 1, females are checked for copulation plugs. Female mice are sacrificed for embryo collection around 10:00 a.m. on the morning of day 1, whereas female rats are sacrificed for embryo collection around 1:00 p.m. on the afternoon of day 1. Embryos are collected from mice and rats using the procedure described in (Nagy et al. 2002), and the chapter by Shirley Pease in this book (Chap. 11, *Ancillary Techniques*).

Animals are sacrificed by CO_2 inhalation; the oviducts are excised and transferred to a dish containing M2 medium at room temperature. Newly ovulated embryos, enclosed by cumulus mass cells, are released from the swollen ampullae (the upper portion of the oviduct) by gently tugging and opening the walls of the ampullae with fine forceps. The embryos are then transferred to a dish containing a hyaluronidase solution (Sigma H 3884, 300 g/ml in M2 medium), which enzymatically digests the cumulus cells, thus releasing the embryos. As soon as the cumulus cells are seen to be shed, usually within several minutes, the embryos are transferred to fresh M2 medium to wash off the hyaluronidase solution and preserve the viability of the embryos. In rats, the cumulus cells adhere tenaciously to the surface of some embryos and are difficult to remove completely, even with hyaluronidase. In general, their presence does not seem to hinder the subsequent manipulation of the zygotes. The embryos are then transferred to microdrops of KSOM medium (Nagy et al. 2002). under mineral oil and cultured in a humidified 37 °C incubator under 5% CO_2 until needed.

1.3.2 Delivery of Lentiviruses to Single-Cell Embryos

Lentiviruses are delivered to the fertilized oocytes on the day of collection, targeting only single-cell zygotes to minimize the likelihood of the occurrence of mosaicism. Two methods are available to deliver lentivirus to embryos.

1.3.2.1 Microinjection of Lentiviruses into the Perivitelline Space of Single-Cell Embryos

Micropipettes are prepared by pulling borosilicate glass capillaries (Sutter, 1 mm O.D., 0.7 mm I.D., no filaments) on a pipette puller (use a program similar to the one used to produce blastocyst injection pipettes). The tip of the pipette is broken with a razor blade to obtain an outer tip diameter of approximately 15 μm. To ease penetration into the zona pellucida, it is advisable that the pipette tip is not blunt-ended, but has a bevel instead. Before loading the pipette with the viral solution, the tip is inspected under a 40× magnification objective in an inverted microscope equipped with a micrometer in one of the oculars. The best pipettes have an outer diameter of between 10 and 20 μm. It is difficult to penetrate the zona pellucida with pipettes larger than 25 μm. It is difficult to load the viral solution into pipettes smaller than 10 μm.

The micropipette is inserted into the pipette holder of a hydraulic injector (Cell Tram Oil or Vario, by Eppendorf is excellent for this application) and set in place on the micromanipulator over the microscope stage. If using an oil-filled line, make sure that no air bubbles are trapped in the line connecting the injector to the micropipette. Air bubbles will make it very difficult to control the flow of the virus when attempting to deliver it into the zygote. A frozen aliquot of lentiviral concentrate is thawed at room temperature and gently pipetted up and down with a 20 μl plastic tip to release large aggregates of cellular debris. The virus is centrifuged at low speed in a table-top microcentrifuge (1,000 rpm for 1 min) to get rid of large aggregates that would certainly clog the injection pipette. To avoid any debris that may have settled at the bottom of the tube, take the virus to be loaded into the pipette from the top of the aliquot. Immediately after spinning, take 4–5 μl viral suspension and place as a drop on a glass coverslip. Put the coverglass on the stage of the inverted microscope and lower the tip of the glass pipette into the drop. The viral solution is pulled up into the micropipette from the tip using gentle negative pressure from the hydraulic device. It usually takes approximately 10 min for 3–4 μl to be loaded into the tip. The loading procedure should be observed under the microscope in order to detect whether cellular debris is clogging the tip of the pipette. It is normal to see debris in the viral supernatant under microscopic examination. However, the particles should be smaller that 2 μm. To prevent clogging, the pipette can be moved around the drop into regions with a lower concentration of debris. If the tip is clogged, apply positive pressure until the plug is expelled, and start again by applying negative pressure. If the pipette clogs repeatedly during loading it is possible that: (1) the pipette tip might be too small, in which case break a bigger tip on a new pipette, or (2) the viral solution might be too dirty. In this case, spin the virus again at low speed to get rid of debris. A 3 μl sample of viral solution should be sufficient to inject some 200 embryos. Try to minimize the time that the virus is at room temperature, as viral half-life under these conditions is quite short: some studies suggest that most of the virus will be inactive after 6 h at room temperature.

Single-cell embryos are transferred to a microdrop of M2 medium on a slide and covered with mineral oil to prevent evaporation and to maintain osmolarity. The slide is mounted on the stage of an inverted microscope and the holding and injection pipettes lowered onto the microinjection slide and the drop containing the embryos, under visual control with the microscope. The microinjection procedure is monitored under 400× magnification. Embryos are held in place against a fire-polished pipette using gentle negative pressure. Gently apply some positive pressure so that the viral concentrate is discharged slowly but continuously from the injection pipette tip. It is easy to detect the flow of virus coming out of the pipette by observing the currents generated in the medium, and by observing the movement of particulate matter in the viral preparation as it is discharged from the injection pipette. Using the micromanipulator to guide the pipette, the tip is pushed in a swift movement, without hesitation, through the zona pellucida into the region between the zona pellucida and the oocyte cell membrane, or oolemma. To prevent rupturing the embryo, it is important not to push the pipette too deep inside the perivitelline space. Under 400× magnification, some cell debris from the viral solution can be seen slowly flowing from the pipette tip into the perivitelline space. If the zona pellucida enlarges substantially when penetrated by the pipette, the flow rate is too high and the viral solution will be depleted before most of the embryos can be injected. Injection of an excessive volume into the perivitelline space may change the shape of the embryo, giving it a concave appearance, but this does not result in the death of the embryo. The tip is left in the perivitelline space for approximately 3 s until the viral preparation can be seen to have drifted one-quarter way around the circumference of the embryo and/or the perivitelline space expands a little, the embryo perhaps becoming crescent-shaped within the zona. Again, this will not result in the death of the embryo. The micropipette is then withdrawn from the zygote. If the virus stops flowing during the injection procedure, apply more positive pressure. If the flow does not resume, the pipette has likely clogged with some large cellular debris. Very slowly, apply positive pressure until the plug can be seen in the pipette tip and, immediately before it appears that the plug is to be expelled from the tip, reduce the pressure to prevent a massive outflow of virus into the medium. If the plug cannot be expelled even by applying high pressure, then the best course of action is to discard the pipette into a biohazard waste container or into a flask with bleach. A new pipette with an appropriate tip must be reloaded to resume injection. After injection, the embryos are sorted – those that are lysed or abnormal are discarded. The remaining embryos are transferred to KSOM microdrops under oil and cultured in a 37 °C incubator under 5% CO_2 until implantation. When embryos are infected by injection through the zona pellucida it is possible to implant them either on the day of injection, or 3 days later when they have reached the blastocyst stage.

1.3.2.2
Co-incubation of Denuded Single-Cell Embryos with Lentivirus

The zona pellucidae of the fertilized oocytes are removed by incubation for several minutes in either an acid Tyrodes solution (Nagy et al. 2002), or a 0.5% pronase solution in M2 medium at 37 °C in a humidified 5% CO_2 incubator. When the zonae appear to be dissolved, embryos are washed in excess M2 medium, then KSOM, to wash away the M2 and then transferred into 10 μl microdrops of viral suspension in KSOM under mineral oil. To prevent the loss of embryos due to adhesion to the glass walls of the transfer pipettes, it is important to pre-coat them with 1% albumin in PBS, or to siliconize them prior to pulling. It may also help to move around only small numbers of embryos at one time, i.e., pick up and move only 10 embryos at a time instead of 100 or more. Embryos are cultured individually in separate microdrops to prevent them from adhering to one another, resulting in the creation of chimeric animals. The viral suspension can be diluted to various concentrations to roughly control the average number of proviral integrations expected per transgenic genome, since the number of insertions should scale down with the concentration of virus used. We usually incubate the zygotes in 10 μl microdrops of virus diluted to 2×10^4 IU/μl, 400 IU/μl, and 80 IU/μl, which result (on average) in seven, three, and one proviral copies per genome. Zygotes are incubated in the viral suspension until the embryos reach the morula stage and are subsequently implanted. As indicated above, denuded embryos may stick to the glass wall of the pipettes, possibly resulting in a significant loss of embryos. Implantation of denuded embryos prior to the morula stage is not possible because embryos with no zona are unable to move down the oviduct towards the uterus. In the implantation of embryos with no zona after incubation with virus, the yield of pups is lower compared with the implantation of injected embryos complete with zona (unpublished observations). Thus it is important to anticipate that, due to the trauma of denudation, survival of embryos in this procedure will be lower as compared to perivitelline injection. Nevertheless, co-incubation of denuded single-cell embryos with lentivirus is a very viable approach for the production of genetically altered mice.

1.3.3 Transfer of Embryos into Recipient Females

The timed mated pseudopregnant females needed to host the treated embryos are prepared by mating sexually mature females in estrus to vasectomized, mature males the night before the intended day of implantation (see Chap. 11, *Ancillary Techniques*, for further options on the timing of these events). Males are vasectomized by tying off the vas deferens at two separate locations approximately 5–6 mm apart, then cauterizing the intervening segment to sever the tube. Males are vasectomized at least 2 weeks prior to mating to ensure that all remaining sperm in the uro-genital tract is dead at the time of mating. For mice, we maintain a large colony of females as possible recipients and, at any given time, some proportion of those females will be receptive to mating, as determined by the presence of a copulatory plug on the morning of the day of intended implantation.

For rats, an alternative to maintaining a large colony of females is to order timed pregnant females that will deliver the day before that of intended implantation. Immediately after delivery, female rats are in estrus and will be receptive to mating that night. Alternatively, vaginal smears can be performed the night before implantation to select the females in estrus to be placed with the vasectomized males. Mating is confirmed the next morning by checking for a copulatory plug.

Embryos infected by perivitelline injection are transferred into host females as soon as possible for maximum rates of implantation. Early stage embryos [0–1 days post coitum (dpc) with an intact zona pellucida are transferred to the oviduct of timed pseudopregnant females (0–1 dpc). Embryos that have reached the morula or blastocyst stage are transferred to the uterus of timed pseudopregnant females (2.5 dpc). In general, no more than 30 embryos are transferred bilaterally into the uterus. These procedures are described in Chap. 11 *Ancillary Techniques*, or in (Nagy et al. 2002). Pregnancy and delivery of the transgenic litter is as usual, although our experience has been that litter sizes from the implantation of lentivirus-injected embryos are generally greater than those from pronuclear-injected embryos (unpublished observation).

1.4 Establishment of Stable Strains from Lentiviral Founder Animals

Each and every transgenic animal developed from an embryo having undergone infection with a lentiviral vector will be genetically different. In animals generated by lentiviral delivery, each viral copy will integrate as a single molecule at a site within the genome, but different copies will integrate independently into different chromosomal locations. This will have two consequences. Firstly, each animal may express the transgene to differing degrees, depending on the integration site and copy number. Secondly, as these founder animals breed, their offspring may carry fewer copies of the transgene, as segregation and independent assortment of chromosomes takes place during gametogenesis in each generation. In order to eliminate this variability, it is necessary to identify animals that express the transgene well and then breed those animals in a manner that will allow a stable strain to be established, each member of which carries one copy of the transgene that is adequately expressed and is inherited in a stable manner.

1.4.1 Mating of Founder Animals

Since founder animals are all different, they should not be mated to each other, but outcrossed or backcrossed to wild-type animals from your strain of choice. Founder animals, i.e., those born from injected embryos, should first be analyzed by Southern hybridization to determine how many copies of the transgene are carried by each founder. To discover the number of copies present in a lentiviral transgenic founder it is necessary to follow two steps. First, digest the genomic DNA with a restriction enzyme that cuts once within the lentiviral construct. In this way, the integrated viral copy will be split into a 5′ side and a 3′ side. The distance between the internal restriction site (in the transgene) to the nearest restriction site in the flanking genomic DNA will determine the size of the band that will be detected by Southern blotting. Second, prepare a probe derived from the lentiviral construct that does not contain the LTRs. If the LTR sequence is included in the probe, it will detect both the 5′-LTR and the 3′-LTR of the integrated proviral transgene. Because the DNA is digested with a restriction enzyme that cuts once inside the viral transgene, a probe that includes the LTRs will detect two bands (one with the 5′-LTR and one with the 3′-LTR) per integrated transgene. Thus, using the appropriate type of probe and digesting the genomic DNA as indicated above, it is possible to precisely count the number of integrated transgene copies. In some cases, however, it is possible that two copies of the transgene could give rise to fragments of almost identical size when the DNA is digested with a particular enzyme. This will occur if, by chance, the restriction sites in the flanking genomic DNA of two independent integrated transgenes are at a similar distance from the internal restriction sites in the transgene. In this case, the two bands revealed by Southern will overlap and it might seem that there is only one band in the Southern blot. In order to prevent this confusion from arising, perform two Southern blots in parallel, digesting the DNA with two different enzymes, one for each blot.

To evaluate expression levels, mate each founder to one or two wild-type mice. Genotype litters (F1 animals) by Southern blot analysis (to identify copy number), and evaluate expression levels by microscopy, western blotting or immunocytochemistry. Animals that have low expression levels of the transgene should be discarded at this step, regardless of the number of copies. Animals that have good expression levels, but have multiple copies of the transgene should be mated again to wild type animals in order to obtain stable lines that show good expression levels with single copies of the transgene.

Regarding transgene segregation, it is important to remember that if two integrated viruses are located on the same chromosome, they will not segregate as easily as they would if they were on different chromosomes.

1.5 Safety Guidelines for Pseudotyped Retroviruses

The biosafety office of your institution must be notified prior to the use of lentiviral vectors, in order to obtain permission and further institution-specific instructions. At a minimum, BL2 conditions should be used at all times when handling the virus.

All decontamination steps should be performed using 70% ethanol and 1% SDS. Gloves should be worn at all times when handling lentiviral preparations, transfected cells, or the combined transfection reagent. Remember that, although the virus has been significantly modified for biosafety, its VSV coat means that human cells can be infected even if they are not dividing. That said, the following modifications have been made to prevent viral replication.

- The packaging genes are separated on two plasmids, which lack LTRs and have no viral packaging signal. For more information regarding lentiviral constructs, please refer to the work of Naldini et al. (Naldini et al. 1996), in which the first generation of recombinant lentiviral vectors is described.
- The following viral genes have been deleted from the packaging vector: *env*, *vpr*, *vpu*, *vif* and *nef*. (Zufferey et al. 1997) describe the use of packaging vectors in which the packaging components have been split onto two separate plasmids to increase biosafety.
- The vector expressing the packaged viral genome has a self-inactivating LTR (TATA box deletion) and expresses no viral gene products. (Miyoshi et al. 1998) describe the procedures necessary to engineer a self-inactivating lentiviral vector.
- Envelope, in this case VSVg, is expressed by a separate vector.

The above modifications make it virtually impossible to accidentally generate a replication competent virus. However, because these viruses are pseudotyped with VSVg envelope, special considerations have to be taken into account.

- Always treat virus solutions, and cells transfected or infected with virus as a potential biohazard. Absolutely always wear gloves. Try to minimize occasions in which tubes are open outside of the tissue culture hood. When injecting the virus into the zygotes be very careful when handling the embryos and the solution in which they are incubated on the microscope stage. Be very careful when disposing of the glass slide that is used to load the virus, the glass capillary used for injection, and the holding pipette.
- Remember that even though you know that you are working with a biohazard, other people who are not aware of it might come into contact with your virus solutions. Do not leave virus-containing solutions unattended: they might be handled by someone who is unaware of the potential risks and thus will not take the necessary precautions. Do not leave virus-containing solutions in the hood or in the centrifuge if you are not there. If your centrifuge run is finished, subsequent users may touch your virus-containing tubes without gloves.
- Decontaminate everything that has been in contact with the virus with a solution containing either 70% ethanol and 1% SDS, or bleach. Wipe the surfaces of the tissue culture hood that you have used. Clean the buckets for the ultracentrifuge thoroughly. Discard the tissue culture plates and other virus-containing plasticware into dedicated biohazard

bags. Aspirate some bleach through the vacuum apparatus to decontaminate the lines. Be especially careful with glass Pasteur pipettes, as they are the main risk of accidental skin breakage.

- Restrict the number of places where the virus is produced and used. Use the viruses only in tissue culture hoods, tissue culture incubators and centrifuges that are labeled with a "Biohazard: pseudotyped viruses" sign. To prevent accidental contamination of items of common use, after using viruses, changes gloves before you touch door handles, controls for equipment, pipettes, etc.
- Do not use pseudotyped viruses encoding genes that are known, or potential, oncogenes in a normal laboratory setting the other researchers use. Furthermore, unless it is completely established that a gene is clearly not oncogenic, do not pseudotype. If you have any doubts, use only in a P3 (BL3) grade level room.

References

Brinster RL, Chen HY, Trumbauer ME, Yagle MK, Palmiter RD (1985) Factors affecting the efficiency of introducing foreign dna into mice by microinjecting eggs. Proc Natl Acad Sci USA 82:4438–4442

Fields-Berry SC, Halliday AL, Cepko CL (1992) A recombinant retrovirus encoding alkaline phosphatase confirms clonal boundary assignment in lineage analysis of murine retina. Proc Natl Acad Sci USA 89:693–697

Gordon JW, Scangos GA, Plotkin DJ, Barbosa JA, Ruddle FH (1980) Genetic transformation of mouse embryos by microinjection of purified dna. Proc Natl Acad Sci USA 77:7380–7384

Hofmann A, Kessler B, Ewerling S, Weppert M, Vogg B, Ludwig H, Stojkovic M, Boelhauve M, Brem G, Wolf E, Pfeifer A (2003) Efficient transgenesis in farm animals by lentiviral vectors. EMBO Rep 4:1054–1060

Hofmann A, Zakhartchenko V, Weppert M, Sebald H, Wenigerkind H, Brem G, Wolf E, Pfeifer A (2004) Generation of transgenic cattle by lentiviral gene transfer into oocytes. Biol Reprod 71:405–409

Ivarie RA (2003) Avian transgenesis: progress towards the promise. Trends Biotechnol 21:14–19

Jaenisch R (1976) Germ line integration and mendelian transmission of the exogenous moloney leukemia virus. Proc Natl Acad Sci USA 73:1260–1264

Jaenisch R (1988) Transgenic animals. Science 240:1468–1474

Jahner D, Stuhlmann H, Stewart CL, Harbers K, Lohler J, Simon I, Jaenisch R (1982) De novo methylation and expression of retroviral genomes during mouse embryogenesis. Nature 298:623–628

Lois C, Hong EJ, Pease S, Brown EJ, Baltimore D (2002) Germline transmission and tissue-specific expression of transgenes delivered by lentiviral vectors. Science 295:868–872

Lois C, Refaeli Y, Qin XF, Van Parijs L (2001) Retroviruses as tools to study the immune system. Curr Opin Immunol 4:496–504

Martinez-Salas E (1999) Internal ribosome entry site biology and its use in expression vectors. Curr Opin Biotechnol 5:458–464

Miyoshi H, Blomer U, Takahashi M, Gage FH, Verma IM (1998) Development of a self inactivating lentivirus vector. J Virol 72:8150–8157

Nagy, A., Gertsenstein, M., Vintersten, K., Behringer, R. (eds) (2002) Manipulating the mouse embryo: a laboratory manual. 3 edn Cold Spring Harbor Laboratory Press, Cold Spring Harbor, NY

Naldini L, Blomer U, Gallay P, Ory D, Mulligan R, Gage FH, Verma IM, Trono D (1996) In vivo gene delivery and stable transduction of non dividing cells by a lentiviral vector. Science 272:263–267

Shimohama S, Rosenberg MB, Fagan AM, Wolff JA, Short MP, Breakefield XO, Friedmann T, Gage FH (1989) Grafting genetically modified cells into the rat brain: characteristics of e. coli beta-galactosidase as a reporter gene. Brain Res Mol Brain Res 5:271–278

Zufferey R, Nagy D, Mandel RJ, Naldini L, Trono D (1997) Multiply attenuated lentiviral vector achieves efficient gene delivery in vivo. Nat Biotechnol 15:871–875

2 Intracytoplasmic Sperm Injection (ICSI) in the Mouse

Ming-Wen Li, K.C. Kent Lloyd

2.1 General Description of Intracytoplasmic Sperm Injection (ICSI)

Intracytoplasmic sperm injection (ICSI) has been developed as an assisted reproductive technique by which a sperm is mechanically injected into the cytoplasm of an oocyte. This technique was pioneered initially in sea urchin (Hiramoto 1962), then in mouse (Lin 1966) and hamster (Uehara and Yanagimachi 1976). Although ICSI has been used successfully in humans throughout the world since it was first introduced clinically in 1992 (Palermo et al. 1992), ICSI in mouse proved extremely difficult, and was unsuccessful until the introduction of the piezo-driven micromanipulator in 1995 (Kimura and Yanagimachi 1995). In theory, ICSI of a single sperm, including sperm DNA and other sperm factors (Wilding and Dale 1997), whether "live" or amotile, is sufficient to derive a viable mouse. In general, ICSI results in a relatively high percentage (> 50%) of successfully fertilized embryos and liveborn offspring, all of which are likely to wean and become reproductively sound. ICSI has been used to derive normal and wildtype mice with an F1 hybrid and/or inbred background, genetically-altered strains (e.g., transgenics, knockouts), strains in which only one male of the genotype remains, and strains in which a male is unable to breed naturally either because it is too old or cannot mate successfully. The genotype and phenotype of a particular strain can be transmitted vertically with high fidelity to the next generation when derived from sperm used for ICSI. ICSI can be used in any situation in which an alternative reproductive technique (ART) is applicable or warranted. In situations in which other ARTs may be applicable, the choice of ICSI should be based on whether it is more advantageous and more likely to result in a successful outcome, i.e., live born offspring, than when using in vitro fertilization (IVF) or artificial insemination (AI). For example, we routinely

Ming-Wen Li: Center for Comparative Medicine, School of Veterinary Medicine, University of California, One Shields Avenue, Davis, CA 95616, USA

K.C. Kent Lloyd: Center for Comparative Medicine, School of Veterinary Medicine, University of California, One Shields Avenue, Davis, CA 95616, USA, e-mail: kclloyd@ucdavis.edu

Principles and Practice
Mammalian and Avian Transgenesis – New Approaches
Shirley Pease and Carlos Lois (Eds.)

use ICSI for resurrecting strains on a C57BL/6 background from cryopreserved sperm, because, in our experience, we are more likely to achieve liveborn offspring from ICSI than from IVF. In other situations, ICSI is the only reasonable choice of ART. For example, ICSI is the only technique that can successfully fertilize oocytes using dead, or amotile sperm. ICSI is also an essential technique when only a few sperm are available, when sperm cannot penetrate the zona pellucidae or oolemma, or when the availability of sperm, such as from a single aged male of a strain, is limited.

2.2 Principles of ICSI

In ICSI, a single sperm, either intact or sperm head only, is mechanically inserted into the oocyte cytoplasm, bypassing all "natural" barriers to fertilization, including the cumulus layer, zona pellucida, and oolemma (plasma membrane). ICSI also precludes the natural incorporation of sperm membrane into the oolemma, as occurs during natural (non-assisted) fertilization. Sperm obtained from the epididymis do not need to be capacitated or acrosome-reacted when used for ICSI, although studies indicate that induction of the acrosome reaction may increase male pronuclear formation after ICSI (Kimura and Yanagimachi 1995).

Mouse sperm has a hooked head that connects through the neck to a long tail, making it rather difficult to use for ICSI. Fortunately, only the sperm head is necessary for ICSI fertilization and embryo development in the mouse (Kuretake et al. 1996). In most mammals, including humans, centrioles and centrosomes located in the neck region are inherited paternally (Schatten 1994). In contrast, mice (and perhaps some other rodents) follow a maternal method of centrosome inheritance and thus the neck region is not essential, which in part explains why ICSI using the sperm head alone is sufficient to obtain normal offspring (Kuretake et al. 1996). A technical advantage of injecting only the sperm head is that the amount of medium co-injected into the oocyte is greatly reduced, which contributes to the overall survival rate of oocytes after ICSI.

After injection into the oocyte cytoplasm, the sperm head loses its nuclear envelope, and its nuclear protein undergoes remodeling, decondenses, and acquires a new nuclear envelope to form the male pronucleus. The male pronucleus becomes functionally competent to synthesize DNA and RNA. The sperm head includes not only DNA but also sperm factors, mostly proteins, essential for egg activation to occur after ICSI fertilization. Spermatocytes and round spermatids do not have the ability to activate the oocyte (Kimura and Yanagimachi 1995; Sasagawa et al. 1998), and thus artificial activation of injected oocytes is required.

2.3 Quality and Treatment of Sperm Used for ICSI

Probably the single most important justification for the development and application of ICSI for mouse colony management is the minimal requirement of sperm amenable to ICSI. Sperm from a male mouse that is fertile, subfertile, or infertile, including a male that either physically cannot mate or cannot produce normal sperm capable of fertilizing oocytes naturally either in vivo or in vitro, can be used successfully for ICSI. Sperm types that have been used successfully include epididymal and testicular sperm, which retains its motility after having been freshly harvested, following complete or hemicastration (Kimura and Yanagimachi 1995), and after having been recovered following cryopreservation ((Wakayama et al. 1998); K.C.K. Lloyd unpublished data). Normal offspring can also be obtained by injection of oocytes with spermatocytes (Kimura and Yanagimachi 1995) and round spermatids (Sasagawa et al. 1998) from testis after artificial activation. It is this latter source that forms the basis for rescue of an infertile line using spermatocytes or round spermatids from a male that cannot complete spermatogenesis (Meng et al. 2002).

In addition, sperm that is inherently poorly motile (e.g., strain related) or that loses motility (Mazur et al. 2000) after recovery following cryopreservation (K.C.K. Lloyd unpublished data; (Wakayama et al. 1998)), freeze-drying (K.C.K. Lloyd unpublished data; (Wakayama and Yanagimachi 1998; Kusakabe et al. 2001)), snap freezing, or evaporative drying, and is thus not conducive for use in either IVF or AI, has been used successfully for ICSI.

This latter fact has, in part, been responsible for the recent popularity of ICSI as a procedure for mouse strain and colony management. For example, very low fertilization rates are achieved after IVF using cryopreserved sperm from certain inbred strains, such as C57BL/6 and BALB/c (K.C.K. Lloyd unpublished data; (Szczygiel et al. 2002)). ICSI of cryopreserved sperm from these two inbred strains achieves high fertilization rates and results in liveborn offspring (K.C.K. Lloyd unpublished data; (Szczygiel et al. 2002)).

Live offspring can also be obtained from ICSI of epididymal sperm obtained from a dead mouse stored in the refrigerator for at least 20 days (Kishikawa et al. 1999a). This is particularly helpful when a valuable male mouse dies unexpectedly and facilities for sperm cryopreservation are not immediately accessible. Sperm can be successfully used for ICSI if obtained from the epididymis and/or vas deferens, or even the cadaver itself after temporary storage at 4 °C.

2.4 ICSI Instrumentation

Despite the very successful and multiple applications of ICSI as an effective ART in mice, the complex nature of the procedure and specialized instrumentation required have been cited by some as making ICSI inaccessible to most workers in the field, with the exception of a few highly specialized laboratories. These are unfair assertions, as the instrumentation differs in only a few respects from that used in most transgenic laboratories today, and the procedure can be mastered with concerted daily practice and effort.

The instrumentation needed to establish ICSI in the laboratory can be divided into two groups: that required for preparation of the microtools, and that required for the ICSI procedure itself. A puller and microforge are necessary to prepare the microtools. The correct settings and use of these instruments are of critical importance in preparing good microinjection pipettes, which in turn is crucial to the ICSI procedure itself. These details are described in Sect. 2.5.

The instrumentation required for the ICSI procedure itself begins with an inverted microscope equipped with the appropriate optics, micromanipulators and microinjectors. Both Nomarski differential interference contrast (DIC) optics and Hoffman modulation contrast (HMC) optics can be used for ICSI. DIC optics are comparatively more expensive than HMC, and require the use of glass as an injection chamber. Hoffman optics are necessary when used with plastic injection chambers, but reportedly provide inferior resolution compared to DIC.

There are currently three main microinjection systems in common use that can be applied to ICSI: Eppendorf, Leitz, and Narishige. Eppendorf sells an electronic micromanipulator using microprocessor-controlled motors. The motion of an electric joystick is translated into a proportional response by three massive stepper motors that drive the microtool holder. Perhaps due to expense and other features, this system has not been applied widely for ICSI. The Leitz micromanipulator is a purely mechanical design based on the eccentric motion of a spherical component that forms the apex of a "hanging joystick". The stability, precision and reliability of the Leitz mechanism has been well documented. However, when used with an inverted microscope, the Leitz micromanipulators must be raised on supports to provide access to the microscope stage, which may make operation of the joystick difficult. The Narishige micromanipulator, available from Nikon, relies on three hydraulic cylinders to drive simultaneous, three-dimensional remote-controlled positioning. The fundamental advantage of the Narishige system is that separate, high-precision micrometer drives are included on both horizontal joystick cylinders. These allow for convenient X-Y motion control making for precise, yet wide-ranging, linear movements.

The ICSI procedure most commonly applied in mice today involves the use of a drilling device to assist in penetration of the zona pellucida and oolemma. ICSI performed without this assistance, referred to as conventional ICSI, is inefficient, and the survival rate of mouse oocytes after conventional ICSI is low (Ron-El et al. 1995; Kimura and Yanagimachi 1995; Yanagimachi 1998). It is for this reason that a drilling device utilizing piezo technology is used instead. The piezo-electric micromanipulator can deliver extremely rapid and high resolution (0.1–0.5 μm) pulses of deformation, which facilitates and simplifies penetration of the zona pellucida and oolemma. The speed and intensity of the piezo controller should be set as low as possible. A small amount of mercury within the tip of the injection pipette can significantly enhance the effectiveness of rapid vibration delivered with each piezo pulse. (NOTE: mercury is a potentially harmful chemical that should be used with caution. Consult your Safety Office regarding precautionary measures to be taken when handling, using and disposing of mercury.) We are most practiced in the use of the piezo-electric device available from Prime Tech (Japan), but similar devices are available from other suppliers, for example, Burleigh's piezo drill.

Finally, because microtools are very sensitive to ambient vibration, conducting ICSI on a commercial anti-vibration table (e.g., TMC-Technical Manufacturing Corporation, Peabody, MA) may be required.

2.5 Preparation of Microtools

The preparation of the holding and microinjection pipettes is of crucial importance to successful ICSI. Both types of pipettes are prepared from borosilicate glass capillaries (O.D. 1 mm, I.D. 0.75 mm, length 10 cm, Sutter Instrument, San Rafael, CA) using a puller. The capillaries are autoclaved or heat sterilized before use. A horizontal puller (e.g. Model P-97, Sutter Instrument Company, Novato, CA) is used to pull the capillaries to form a pipette with a long, fine, and straight tip. Different parameters, including pressure, temperature, time, pulling force and velocity, influence the shape of the needle. On the Model P-97 horizontal puller, using the Sutter capillaries described above, the relative settings we have been using for each of these variables are as follows: pressure "500", heat "775", pull "65", velocity "130", and time "200" (equivalent to 100 ms).

Pipettes are finished on a microforge (e.g., Narishige MF-900). For making a microinjection pipette, the glass capillary is mounted horizontally on the microforge, and is moved until the outer diameter at the level of the filament is 7–8 μm. A small amount of residual glass ("glass node") on the heating element is heated gently, and then moved towards, and briefly placed in contact with, the capillary. The heating element is turned off after contact with the capillary, which breaks in the appropriate place as it cools. The tip of this newly fashioned microinjection pipette is removed immediately from the glass node after it breaks to protect it from being polished by the remaining heat of the glass node. The microinjection pipette should have an internal diameter of approximately 6 μm and will be ready for use after breaking as described here.

To make a holding pipette, the glass capillary is mounted horizontally on the microforge, and is moved until its outer diameter at the level of the filament is 60–80 μm. The glass node is heated and moved toward the pipette until it touches it. Heating is continued until the pipette softens. Heating is then turned off suddenly, and the pipette breaks in the appropriate place as it cools. The pipette is then changed to a vertical position, and the tip is brought close to, but not in contact with, the heating element. The filament is heated to melt the tip until its internal diameter is between 10 and 15 μm. Using the heating element, the holding pipette is then bent to an angle of about 15°. The finished holding and microinjection pipettes should be stored in clean, dust-free, glass or plastic containers.

2.6 Culture Media

Either M2 medium (Specialty Media, Phillipsburg, NJ) or Hepes-CZB medium (Specialty Media) can be used for oocyte collection (Kimura and Yanagimachi 1995). Polyvinyl pyrrolidone (PVP; 10%, molecular weight approximately 360 kDa, Sigma, St. Louis, MO) in Hepes-CZB is used to slow the movement of the sperm, and to aid control of sperm movement within the microinjection pipette. For microinjection of the sperm head into the oocyte, Hepes-CZB supplemented with 0.01% polyvinyl alcohol (PVA, cold water soluble, molecular weight average 30–70 kDa) without bovine serum albumin (BSA) is used. CZB medium (Specialty Media; (Chatot et al. 1989, 1990)) supplemented with 5.56 mM D-glucose or KSOM medium (Specialty Media) plus amino acids and glucose (Summers et al. 2000) is used for culture of oocytes and ICSI-fertilized embryos. Media for use in holding and working with embryos outside a CO_2 incubator should be buffered with Hepes, in order to maintain the correct pH for optimal embryo survival and development. Media that is not buffered with Hepes, but instead with bicarbonate, is for use in the culture of embryos in a CO_2 incubator. Thus, of the different media mentioned here, CZB and KSOM are used in culture of embryos within incubators. To maintain an appropriate pH and temperature, bicarbonate-buffered CZB or KSOM medium microdrops covered with mineral oil (embryo tested) should be preequilibrated in a humidified 5% CO_2 incubator for a few hours or overnight before use.

2.7 ICSI Procedure

2.7.1 Preparation of Sperm

1. Sacrifice male mouse by cervical dislocation or CO_2 asphyxiation. For collection of epididymal sperm, mice should be between 6 and 8 weeks of age or older.

2. Place mouse in dorsal recumbency on a paper towel and wipe the urogenital region with 70% ethanol. Cut the skin with dissection scissors at a point level with the top of the legs. Grasp the skin firmly on both sides of the incision and pull the skin outward toward the head and tail to expose the body wall. Cut open the body wall to expose the abdominal cavity. Identify and dissect out the two caudal epididymides and place them in Hepes-CZB plus 0.01% PVA (without BSA), or in a suitable cryopreservation solution if sperm is to be cryopreserved.

3. Release the sperm from the caudal epididymis by gentle pressing and/or squeezing with two 30-gauge needles under a dissecting microscope. Disperse the sperm mass by gentle hand-shaking for 1–2 s and warming at 37 °C for 2–3 min.

4. Wash sperm twice to get rid of epididymal secretion by centrifugation (300 g for 2 min) and resuspension (in 1 ml Hepes-CZB). Afterwards, the sperm pellet should be resuspended in 100–200 μl Hepes-CZB plus 0.01% PVA (without BSA) until use.

2.7.2 Preparation of Oocytes

1. Oocytes are collected after superovulation. Superovulation is induced by intraperitoneal (ip) injection of 5 IU pregnant mare serum gonadotropin (PMSG) followed by 5 IU of human chorionic gonadotropin (hCG) 47–48 h later. The best age of females for superovulation varies from strain to strain, but usually falls between 3 and 5 weeks during the

prepubescent stage of development. For example, the optimal age for C57BL/6 J females is 25 days and for BALB/c mice is 21 days. In general, 3- to 8-week-old females are suitable for superovulation.

2. Sacrifice female mice by cervical dislocation or CO_2 asphyxiation 13–14 h after hCG injection.

3. Collect oviducts into warm M2 medium and release the oocyte-cumulus complexes by tearing open the swollen ampulla using a fine needle.

4. Remove cumulus complexes by treatment with 300 U/ml bovine testis hyaluronidase in Hepes-CZB or M2 medium for 3–5 min. The oocytes in hyaluronidase solution should be watched carefully and removed and washed as soon as the cumulus cells have dissociated.

5. Rinse oocytes three times in equilibrated CZB or KSOM medium and incubate in final rinse at 37 °C in 5% CO_2 before ICSI. Oocytes for ICSI should be used within 3–4 h.

2.7.3 Injection Procedure

1. Microtool set-up: holding and microinjection pipettes are prepared from borosilicate glass capillaries using a puller and microforge (see Sect. 2.5). Place approximately 3–5 mm mercury in the proximal end of the microinjection pipette and push the mercury to the tip of the pipette using oil or distilled water from a syringe. Blow out air bubbles from the microinjection pipette holder attached to the drive unit of a piezo-combined micromanipulator, and install the microinjection pipette into the holder. The drive unit of the piezo-micromanipulator is driven by a controller. Place the microinjection pipette holder into the groove of a universal joint attached to a micromanipulator, loosely fix it and adjust the angle of the microinjection pipette to as horizontal as possible. The angle of the injection needle may affect embryo survival after the physical trauma of injection. If the injection pipette can be set at an angle of 15–20° against the injection dish, then a straight injection pipette may be used successfully. If this is not possible, then an improvement may be to put a bend in the injection pipette, such that the angle of penetration of embryos is more horizontal. In this case, the elbow should be far away from the tip of the pipette where the mercury is located. If the elbow is in the region of the mercury, then it will affect the flow control of the fluid and reduce the mercury-enhanced piezo vibration. Fill the holding pipette with mineral oil and install it into the holding pipette

holder before placing the holder into the groove of another universal joint attached to another micromanipulator, and fix it in place.

2. Preparation of microinjection chamber: the cover of a plastic (100 mm×15 mm, Falcon) or glass (Glaswerk, Wertheim, Germany) Petri dish is used as a microinjection chamber. Place a row of medium drops (2–5 μl 10% PVP in Hepes-CZB medium without BSA) along the center of the dish. The first drop is for washing the microinjection pipette, the second is for sperm suspension, and the third is for washing the sperm. The remaining 1–2 drops in the row (10–20 μl Hepes-CZB containing 0.1 mg/ml PVA without BSA) are the microinjection drops. To inject more oocytes in the same chamber, a few more rows can be prepared. Cover the drops immediately with mineral oil.
3. Transfer of sperm to PVP drop: transfer a little sperm suspension with a mouth piece pipette into the second PVP drop and mix it thoroughly. The sperm suspension drop can also be prepared by placing 2–5 μl sperm suspension in PVP onto the chamber before covering with oil.
4. Cutting the sperm head: wash the microinjection pipette in the first PVP drop, and then move the pipette to the second PVP drop containing sperm. Pick up either a motile or an amotile sperm as long as it appears morphologically normal, tail first, into the microinjection pipette. Move the sperm back and forth until the head-midpiece junction (the neck) is at the opening of the microinjection pipette. Apply one or more piezo pulses (intensity 4–5, speed 1–2. PMM controller, Prime Tech, Japan) to the neck region to separate the sperm head from the tail. Do not touch the sperm head when applying piezo pulses. 5–10 sperm heads are prepared in 1–3 min for injection of one group of oocytes.
5. Washing sperm heads: pick up the sperm heads into the microinjection pipette, and move the pipette into the third PVP drop. Expel the sperm heads out into the medium, mix with surrounding medium, and then aspirate them back into the microinjection pipette one by one with a small amount of medium interspersed between each sperm head. Sperm heads should be injected immediately after 5–10 sperm heads are prepared.
6. Sperm head injection: move the microinjection pipette into an injection drop. Transfer 5–10 mature oocytes into the injection drop after washing in Hepes-CZB. Mature oocytes are identified by extrusion of a first polar body. Hold an oocyte with the holding pipette at the 9 o'clock position and the metaphase II spindle (as indicated by a hump in the oocyte cortex) at either the 12 or the 6 o'clock position (Fig. 2.1). Gently aspirate the zona pellucida at the 3 o'clock position using the microinjection

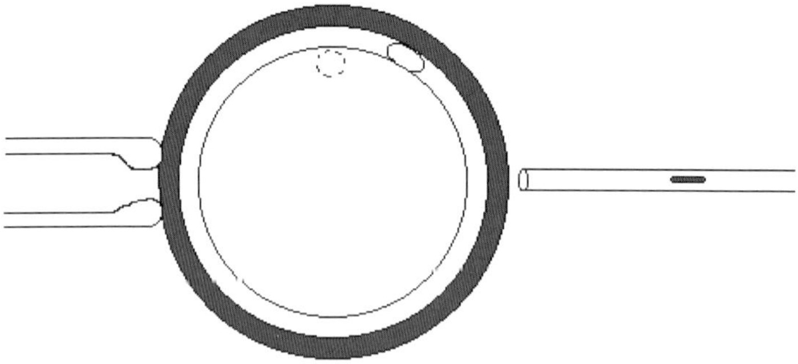

Fig. 2.1. Placement of oocyte. An oocyte is held at the 9 o'clock position by a holding pipette, and the spindle is located at the 12 o'clock position. The microinjection pipette containing a sperm head is ready for penetration at the 3 o'clock position

pipette, and apply several piezo- pulses (intensity 2–4, speed 2–3, for the Prima Tech piezo device) to penetrate the zona pellucida (Fig. 2.2). Expel the piece of the zona pellucida from the microinjection pipette into the perivitelline space. Move the sperm head forward, and advance the tip of the pipette against the oolemma to the opposite side of the oocyte cortex (Fig. 2.3). Penetrate the oolemma by applying one or two piezo-pulses (intensity 1–2, speed 1–2), and expel one sperm head into the ooplasm with a minimum amount of medium (Fig. 2.4). Withdraw the pipette gently (Fig. 2.5). Finish injection of 5–10 oocytes within 10–15 min after sperm heads are ready for injection.

7. After injection: rinse the injected oocytes with CZB or KSOM medium and incubate at 37 °C in 5% CO_2 for a few hours (1-cell stage) or overnight (2-cell stage) before transferring them into the oviducts of pseudopregnant recipient mice. Zona piece may be expelled into the perivitelline space.

2.7.4 Troubleshooting

A sticky sperm preparation can significantly increase the difficulty and time required for preparation of the sperm heads. To avoid this, use only epididymal sperm, if available. Otherwise, other sperm (from the vas deferens) can be used, but be prepared that it may be stickier. If needed, sticky sperm can be washed in Hepes-buffered CZB medium once or twice by centrifugation and resuspension before being used for ICSI.

If the sperm head cannot easily be cut off, the piezo may not be working well. Possible causes include:

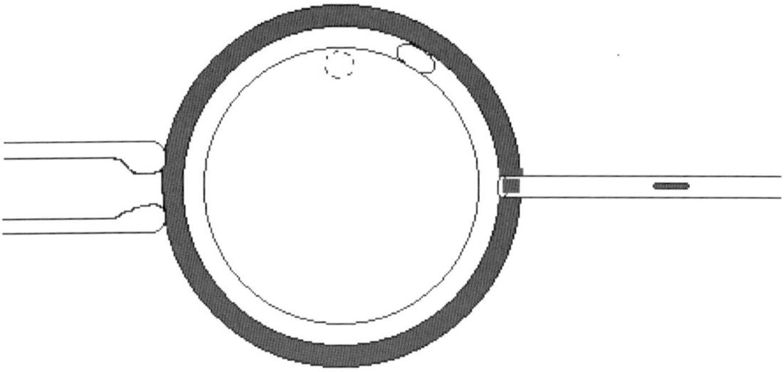

Fig. 2.2. Penetration of zona pellucida. The piezo-driven microinjection pipette has penetrated through the zona pellucida after applying a few piezo pulses; notice the drilled section of the zona pellucida is now inside the tip of the microinjection pipette tip

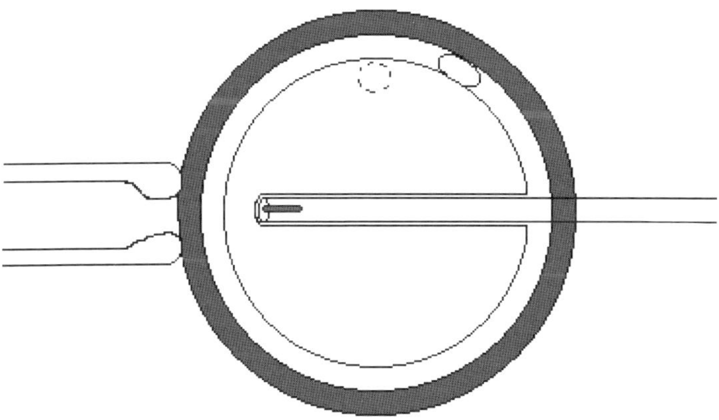

Fig. 2.3. Advance the tip of the pipette. After the section of zona is expelled into the perivitelline space, the microinjection pipette is pushed against the oolemma to the other side of the oocyte, and the sperm head is pushed to the tip

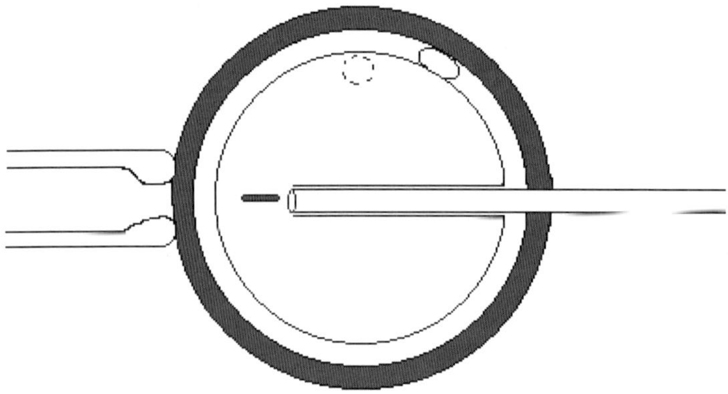

Fig. 2.4. Expel sperm head. The oolema is broken by a single piezo pulse, and the sperm head is injected into the cytoplasm

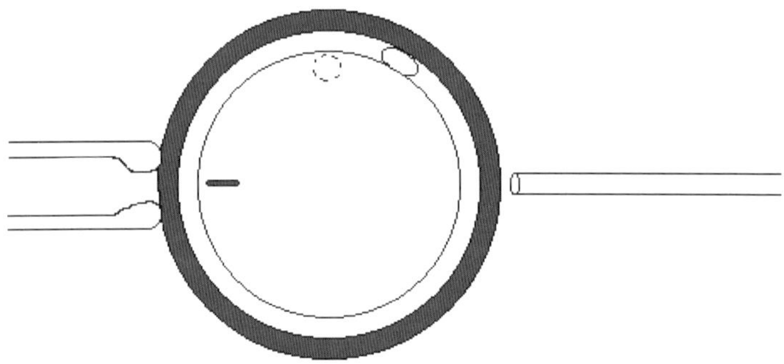

Fig. 2.5. Withdrawal of pipette. The microinjection pipette is withdrawn gently from the oocyte, leaving behind the sperm head

1. the mercury is too far away from the tip opening. Improvement can be made by moving it closer to the tip by means of the microinjector.

2. there may be air bubbles present in the PVP medium in front of the mercury or within the oil- or water-filled microinjection pipette. Correct by pushing bubbles out using the microinjector, or replace the injection pipette.

3. the inside wall of the injection pipette is not smooth, so debris or oil is sticking to the inside wall of the pipette in the mercury region or in front of the mercury. Blowing out some mercury from the injection pipette into the PVP washing drop can help remove debris and oil, so that the mercury can move freely inside the pipette. Mercury discarded in this way must be removed from the PVP drop before the dish is discarded, or the whole injection dish must be disposed of as hazardous material.

An antivibration table is helpful in avoiding interference from building or mechanical vibration transmitted through the floor to the microscope.

Poor fertilization and/or embryo survival rates may be due to too large a volume of medium being injected into the oocyte upon injection of the sperm head. Also, too much physical damage to the oolemma or cytoplasm will cause the embryo to lyse.

2.7.5 Embryo Transfer

Embryos from the one-cell stage through to the blastocyst stage can be transferred into the oviduct or uterus of a pseudopregnant recipient to complete their development. Embryos from the one-cell through the morula stage are transferred into the oviduct of 0.5-days post coitum (dpc) recipients, whereas blastocysts are transferred into the uterine horns of a 2.5-dpc recipient. We have also successfully transferred blastocysts into the oviducts of a 0.5-dpc recipient. Usually, embryos derived by ICSI are transferred at the two-cell stage.

2.7.5.1 Preparation of Pseudopregnant Recipient (Foster) Female Mice

1. Day 0, 6:00 p.m. Mate 6- to 8-week-old CD1 female mice with sterile vasectomized male mice.

2. Day 1, 8:00 a.m. Check for copulation plugs; females ready to serve as recipients of one-cell to blastocyst stage embryos after 12:00 noon.

2.7.5.2 Oviduct Transfer Procedure

1. Sterilize all instruments by dipping them in 100% ethanol and air-drying. Instruments can also be sterilized by autoclaving or by dry sterilizer (e.g., Germinator 500 dry sterilizer).

2. Weigh, then anesthetize pseudopregnant females by ip injection of 2.5% Avertin (0.015–0.017 ml per gram body weight). A stock of 100% Avertin prepared by mixing 10 g tribromoethyl alcohol with 10 ml tertiary amyl alcohol can be stored wrapped in foil at 4 °C. To use, dilute to 2.5% in 0.9% NaCl. After filter sterilization, store wrapped in foil at 4 °C.

3. Place a piece of paper towel on the stage of a dissecting microscope, and place the mouse on the paper towel.

4. Shave the lower back of the mouse and disinfect the dorso-lateral abdomen with 1% povidone iodine and 70% ethanol.

5. Transfer embryos from the CO_2 incubator to warm M2 medium under a dissecting microscope.

6. Load a transfer pipette with 6–10, two-cell embryos, (or 5–6 blastocysts). The narrow part of the pipette should be 2–3 cm in length and 120–180 μm in diameter. Fill the transfer pipette with M2 medium or light mineral oil to just past the shoulder. Take a small air bubble up into the transfer pipette, then M2 medium followed by a second air bubble. Draw up the embryos in a minimal volume of M2 medium and take up a third air bubble, followed by short column of M2 medium. Store the transfer pipette until ready to perform the oviduct transfer.

7. Make a small longitudinal incision (about 1 cm) along the midline at the thoraco-lumber junction with fine dissection scissors. Slide the skin to the left or right to facilitate access for the deeper incision at the level of the last rib.

8. Identify the location of the ovary (orange) or fat pad (white) associated with the ovary, both of which are visible through the body wall. Then make a small incision just over the ovary with fine dissection scissors.

9. Use blunt forceps to pick up the fat pad and gently pull out the ovary, oviduct and part of uterus. Clip a Serrefine clamp onto the fat pad and

lay it down on the back so that the oviduct and ovary remain outside the body wall.

10. Puncture a hole in the ovarian bursa and locate the infundibulum.
11. Hold the edge of the infundibulum with fine forceps and insert the tip of the prepared transfer pipette. Blow on the transfer pipette until the second and third air bubbles have entered the ampulla.
12. Unclip the Serrefine clamp and return the uterus, oviduct, ovary and ovarian fat pad to the abdominal cavity using blunt fine forceps, carefully confirming that all structures are intraperitoneal.
13. Repeat steps 4–11 to transfer additional embryos to the other oviduct, (6–10 two-cell embryos or 5–6 blastocysts), if more embryos are available. Close skin incision in one layer with one or two surgical staples.
14. Keep the mouse warm (37 °C) on a heating pad until it has fully recovered from anesthesia. Afterwards, return the mouse to its cage and its home room.

2.8 Transgenesis by ICSI

Using sperm as a vehicle for delivering exogenous DNA into an egg is an interesting potential application for ICSI in the field of animal transgenesis and biotechnology. The ability of sperm to take up and bind exogenous DNA and then transfer it into an egg during fertilization was first described 30 years ago (Brackett et al. 1971) and was rediscovered in 1989 (Lavitrano et al. 1989). Studies have shown that mouse epididymal sperm can take up $1.5–4 \times 10^6$ DNA molecules (average 5 kB). The mechanism of DNA uptake and binding of sperm relies on an ionic interaction that takes place in the subacrosomal segment of the sperm head between DNA and a group of positively charged sperm membrane proteins. Sperm-bound DNA is rapidly internalized into nuclei in 60–65% of sperm (Lavitrano et al. 1992; Francolini et al. 1993; Zani et al. 1995). While the usefulness of this method applied to sperm to be used for transgenesis by IVF remains controversial, exogenous DNA can be delivered successfully into an oocyte by coinjection with sperm, suggesting an adaptable method of transgenesis (Perry et al. 1999). Clearly, further studies in this area are required to refine and improve these methods in order to establish them as a viable and reliable tool for deriving transgenic mice.

2.9 Reasonable Cautions and Concerns Regarding ICSI

ICSI is a remarkably effective ART for male mice with infertility or sperm dysfunction/dysmotility due to any cause. However, by injecting a single sperm directly into the ooplasm of oocytes, the ICSI technique bypasses all natural sperm selection processes (cumulus penetration, zona binding, acrosome reaction, zona pellucida penetration, oolemma fusion, etc.). Consequently, the delivery of genetic material into an oocyte during ICSI raises concerns of contamination with pathogens and subsequent infection of the embryo.

Another major concern is that of the genetic integrity of the sperm DNA. It is recommended that, whenever possible, motile sperm with normal morphology be used for ICSI. If the sperm has apparently normal morphology and is motile, there is a good chance of successful fertilization and embryo development. If the sperm is of very poor morphology or is largely amotile, then selection of the "best" sperm for ICSI becomes highly subjective. Structurally abnormal sperm are likely to have chromosome abnormalities (Kishikawa et al. 1999b). If motile sperm are not available, then amotile sperm with normal morphology (especially with normal head morphology) should be selected, although normal mice can develop from oocytes injected with sperm with grossly misshapen heads (Burruel et al. 1996). For all of these reasons, ICSI should be considered only if IVF is unlikely to be more successful, or if IVF is not applicable.

Acknowledgements. This work was supported in part by NIH UO1 HD38205 (Lloyd) and by NIH U42 RR14905 (Barthold and Lloyd), which supports the Mutant Mouse Regional Resource Center (MMRRC) at UC Davis, and the UC Davis Mouse Biology Program (http://ccm.ucdavis.edu/mbp).

References

Brackett BG, Boranska W, Sawicki W, Koprowski H (1971) Uptake of heterologous genome by mammalian spermatozoa and its transfer to ova through fertilization. Proc Natl Acad Sci USA 68:353–357

Burruel VR, Yanagimachi R, Whitten WK (1996) Normal mice develop from oocytes injected with spermatozoa with grossly misshapen heads. Biol Reprod 55:709–714

Chatot CL, Lewis JL, Torres I, Ziomek CA (1990) Development of 1-cell embryos from different strains of mice in CZB medium. Biol Reprod 42:432–440

Chatot CL, Ziomek CA, Bavister BD, Lewis JL, Torres I (1989) An improved culture medium supports development of random-bred 1-cell mouse embryos in vitro. J Reprod Fertil 86:679–688

Francolini M, Lavitrano M, Lora Lamia French D, Frati L, Cotelli F, Spadafora C (1993) Evidence for nuclear internalization of exogenous DNA into mammalian sperm cells. Mol Reprod Dev 34:133–139

Hiramoto Y (1962) Microinjection of live spermatozoa into sea urchin eggs. Exp Cell Res 27:416

Kimura Y, Yanagimachi R (1995) Development of normal mice from oocytes injected with secondary spermatocyte nuclei. Biol Reprod 53:855–862

Kishikawa H, Tateno H, Yanagimachi R (1999a) Fertility of mouse spermatozoa retrieved from cadavers maintained at 4 °C. J Reprod Fertil 116:217–222

Kishikawa H, Tateno H, Yanagimachi R (1999b) Chromosome analysis of BALB/c mouse spermatozoa with normal and abnormal head morphology. Biol Reprod 61:809–812

Kuretake S, Kimura Y, Hoshi K, Yanagimachi R (1996) Fertilization and development of mouse oocytes injected with isolated sperm heads. Biol Reprod 55:789–795

Kusakabe H, Szczygiel MA, Whittingham DG, Yanagimachi R (2001) Maintenance of genetic integrity in frozen and freeze-dried mouse spermatozoa. Proc Natl Acad Sci USA 98:13501–13506

Lavitrano M, Camaioni A, Fazio VM, Dolci S, Farace MG, Spadafora C (1989) Sperm cells as vectors for introducing foreign DNA into eggs: genetic transformation of mice. Cell 57:717–723

Lavitrano M, French D, Zani M, Frati L, Spadafora C (1992) The interaction between exogenous DNA and sperm cells. Mol Reprod Dev 31:161–169

Lin TP (1966) Microinjection of mouse eggs. Science 151:333

Mazur P, Katkov II, Katkova N, Critser JK (2000) The enhancement of the ability of mouse sperm to survive freezing and thawing by the use of high concentrations of glycerol and the presence of an escherichia coli membrane preparation (oxyrase) to lower the oxygen concentration. Cryobiology 40:187–209

Meng X, Akutsu H, Schoene K, Reifsteck C, Fox EP, Olson S, Sariola H, Yanagimachi R, Baetscher M (2002) Transgene insertion induced dominant male sterility and rescue of male fertility using round spermatid injection. Biol Reprod 66:726–734

Palermo G, Joris H, Derde MP, Van Steirteghem AC (1992) Pregnancies after intracytoplasmic injection of single spermatozoon into an oocyte. Lancet 340:17–18

Perry ACF, Wakayama T, Kishikawa H, Kasai T, Okabe M, Toyoda Y, R Yanagimachi (1999) Mammalian transgenesis by intracytoplasmic sperm injection. Science 284:1–4

Ron-El R, Liu J, Nagy E (1995) Intracytoplasmic sperm injection in the mouse. Hum Reprod 10:2831–2834

Sasagawa I, Tateno T, Adachi Y, Kubota Y, Nakada T (1998) Round spermatids from prepubertal mouse testis can develop into normal offspring. J Androl 19:196–200

Schatten G (1994) The centrosome and its inheritance: the reduction of the centrosome during gametogenesis and its restoration during fertilization. Dev Biol 165:299–335

Summers MC, McGinnis LK, Lawitts JA, Raffin M, Biggers JD (2000) IVF of mouse ova in a simplex optimized medium supplemented with amino acids. Hum Reprod 15:1791–1801

Szczygiel MA, Kusakabe H, Yanagimachi R, Whittingham DG (2002) Separation of motile populations of spermatozoa prior to freezing is beneficial for subsequent fertilization in vitro: a study with various mouse strains. Biol Reprod 67:287–292

Uehara T, Yanagimachi R (1976) Microsurgical injection of spermatozoa into hamster eggs with subsequent transformation of sperm nuclei into male pronuclei. Biol Reprod 15:467–470

Wakayama T, Whittingham DG, Yanagimachi R (1998) Production of normal offspring from mouse oocytes injected with spermatozoa cryopreserved with or without cryoprotection. J Reprod Fertil 112:11–17

Wakayama T, Yanagimachi R (1998) Development of normal mice from oocytes injected with freeze-dried spermatozoa. Nat Biotechnol 16:639–641

Wilding M, Dale B (1997) Sperm factor: what is it and what does it do? Mol Hum Reprod 3:269–273

Yanagimachi R (1998) Intracytoplasmic sperm injection experiments using the mouse as a model. Hum Reprod 13 [Suppl 1]:87–98

Zani M, Lavitrano M, French D, Lulli V, Maione B, Sperandio S, Spadafora C (1995) The mechanism of binding of exogenous DNA to sperm cells: factors controlling the DNA uptake. Exp Cell Res 217:57–64

3 Generation of Embryonic Stem (ES) Cell-Derived Embryos and Mice by Tetraploid-Embryo Complementation

Kevin Eggan, Rudolf Jaenisch

3.1 Introduction

The generation of embryonic stem (ES)-cell-derived embryos by introducing ES cells into tetraploid embryos is a powerful method for testing the developmental potential of ES cells and for genetic analysis in the laboratory mouse. Production of these embryos, which is often termed tetraploid-embryo complementation, involves (1) the production of tetraploid embryos by electrofusion at the two-cell stage, (2) injection of tetraploid blastocysts with ES cells, and finally (3), transfer of these composite embryos into a suitable surrogate mother (Fig. 3.1). In these chimeric embryos, the extraembryonic lineages are derived completely from the tetraploid host embryo, while the embryonic lineages are comprised entirely of the differentiating derivatives of the ES cells (Nagy et al. 1990).

Because of the compartmentalization of the host and ES cell derivatives, tetraploid embryo complementation is a particularly useful technique for establishing the relative importance of a gene's function in embryonic or extraembryonic development (Rossant and Cross 2001). Whether a mutant phenotype is rescued by normal trophoblast tissues derived from the tetraploid component of the conceptus can be tested by aggregating mutant embryos with tetraploid embryos or by injecting mutant ES cells into tetraploid blastocysts.

Mice completely derived from wild-type ES cells by tetraploid embryo complementation (ES cell-tetraploid) survive to birth at a high frequency, consistent with the notion that ES cells have the developmental potential to give rise to all embryonic lineages (Nagy et al. 1990). However, a problem with routine production of mice using this procedure was that neonatal ES cell-tetraploid mice died at birth due to respiratory failure (Nagy et al. 1990). Since the first description of this technique, a number of ES cell lines

Kevin Eggan: The Stowers Medical Institute, Department of Molecular and Cellular Biology, Harvard University, Cambridge, MA 02138, USA, e-mail: Eggan@mcb.harvard.edu

Rudolf Jaenisch: Whitehead Institute for Biomedical Research and Department of Biology, Massachusetts Institute of Technology,
9 Cambridge Center, Cambridge, MA 02142, USA

Principles and Practice
Mammalian and Avian Transgenesis – New Approaches
Shirley Pease and Carlos Lois (Eds.)

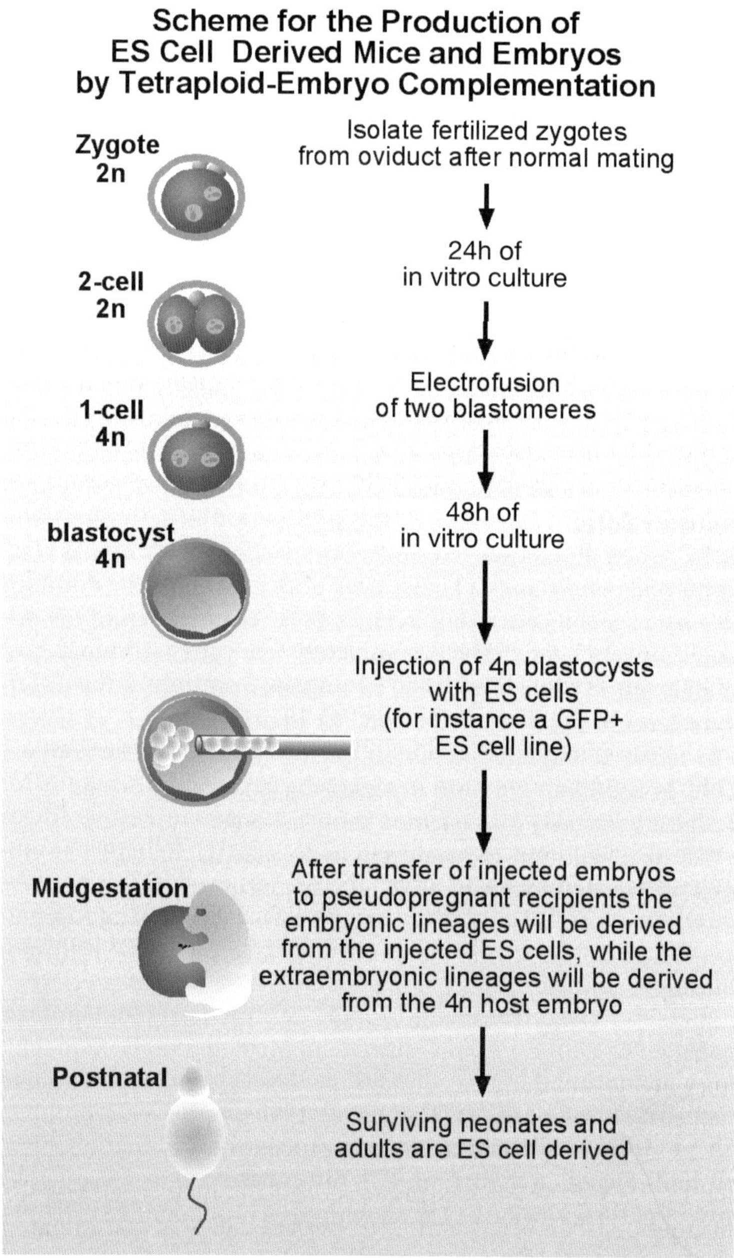

Fig. 3.1. Scheme for the production of embryonic stem (ES)-cell-derived mice and embryos by tetraploid embryo complementation

have been derived or identified that allow the production of mice that survive to adulthood, though at a low frequency (Nagy et al. 1993; Wang et al. 1997). However, the properties of these ES cell lines that permitted neonatal survival remained unclear. Furthermore, the efficiency by which viable mice could be produced, particularly after long-term in vitro culture, was still limited. Thus, the use of tetraploid embryo complementation for the production of mice from targeted ES cell lines did not become widespread due to its unpredictability and inefficiency.

Recently, we have established that hybrid vigor plays an important role in the survival of ES-derived animals (Eggan et al. 2001). Animals produced from ES cells with an inbred genetic background die of respiratory failure with a high penetrance (Table 3.1) (Nagy et al. 1990; Eggan et al. 2001). In contrast, tetraploid embryo complementation using a number of ES cell lines with an F1 genetic background, (i.e., ES cell lines derived from embryos produced by intercrossing parents with different inbred genetic backgrounds), results in the efficient production of viable and fertile animals (Table 3.2) (Eggan et al. 2001). Remarkably, several of these "F1 ES cell lines" retain the potency to efficiently generate viable mice even after as many as five serial rounds of in vitro gene targeting (Eggan et al. 2002). Therefore, tetraploid embryo complementation with F1 ES cell lines represents a reasonable and reliable strategy for the production of embryos and mice from genetically engineered ES cells.

Table 3.1. Development and survival of inbred embryonic stem (ES) cell-tetraploid pubs

ES cell line	Genotype	4 N blasts injected	Pups alive at term (% injected)	Pups respirating after C-section (% alive)	Pups surviving to adulthood (% alive)
J1	129/Sv	120	9(7.5)	0	0
V18.6	129/Sv	48	5(10)	1(20)	0
V26.2	C57BL/6	72	3(4)	1(33)	0
V39.7	BALB/c	72	3(4)	1(33)	1(33)
Total	Inbred	312	20(6)	3(15)	1(5)

In this chapter we will describe strategies used in our laboratories for the direct production of embryos and mice from genetically manipulated ES cells. Our goal is to provide sufficient detail of experimental approaches and methodologies, such that the reader will be able to use the information garnered here to accelerate genetic analysis in their laboratory. With these objectives in mind, we have written protocols that assume a general

Table 3.2. Development and survival of F1 ES cell-tetraploid pups

ES cell line	Genotype	4 N blasts injected	Pups alive at term (% injected)	Pups respirating after C-section (% alive)	Pups surviving to adulthood (% alive)
V6.5	C57BL/6 × 129/Sv	72	18(25)	17(94)	16(89)
V6.5[a]	C57BL/6 × 129/Sv	60	11(18)	9(81)	9(81)
V6.5[b]	C57BL/6 × 129/Sv	20	1(15)	1(100)	1(100)
129B6	129/Sv×C57BL/6	48	2(4)	1(50)	1(50)
F1.2–3	129/Sv×M. Cast.	48	4(8)	3(75)	3(75)
V8.1	129/Sv×FVB	24	7(30)	7(100)	7(100)
V17.2	BALB/c×129/Sv	48	13(27)	12(92)	11(85)
V30.11	C57BL/6×BALB/c	24	4(30)	4(100)	3(75)
Total	F1	344	60(18)	54(90)	51(85)

[a] Three ES cell subclone targeted at the Rosa26 locus

[b] ES cell subclone serially targeted once at the Rosa26 locus and once with a random insertion

knowledge of methods routinely used for the production of mutant and transgenic mice from ES cells. These methods include targeted mutagenesis of ES cell by homologous recombination, and injection of ES cells into diploid blastocysts for the creation of chimeric offspring. For those not well versed in these arts, and as a companion to this chapter, we recommend the other chapters of this book and the text by (Hogan et al. 1994), which has recently been revised by Nagy and his colleagues (Nagy et al. 2002). Additional reviews and information concerning tetraploid embryo complementation and the production of tetraploid embryos can also be found elsewhere ((Rossant and Cross 2001; Mclaughlin 1993; Eggan and Jaenisch 2003); http://www.mshri.on.ca/nagy).

3.1.1 Advantages of Using Tetraploid Embryo Complementation

The production of genetically engineered mice by conventional methods is highly time consuming, and involves the generation of mutant alleles in ES cells by homologous recombination, the production and breeding of chimeric founder mice and, finally, crosses between the resulting heterozygous mice to produce homozygous mutant offspring (Fig. 3.2A). The primary advantage of using tetraploid-embryo complementation in the production of mutant mice is that the heterozygous mice can be produced directly from ES cells without a chimeric intermediate (Nagy et al. 1993; Wang et al. 1997; Eggan et al. 2001). In addition, experimental mice carrying multiple mutations or transgenes may be produced directly from engineered ES cells without breeding (Figs. 3.2B, 3.3).

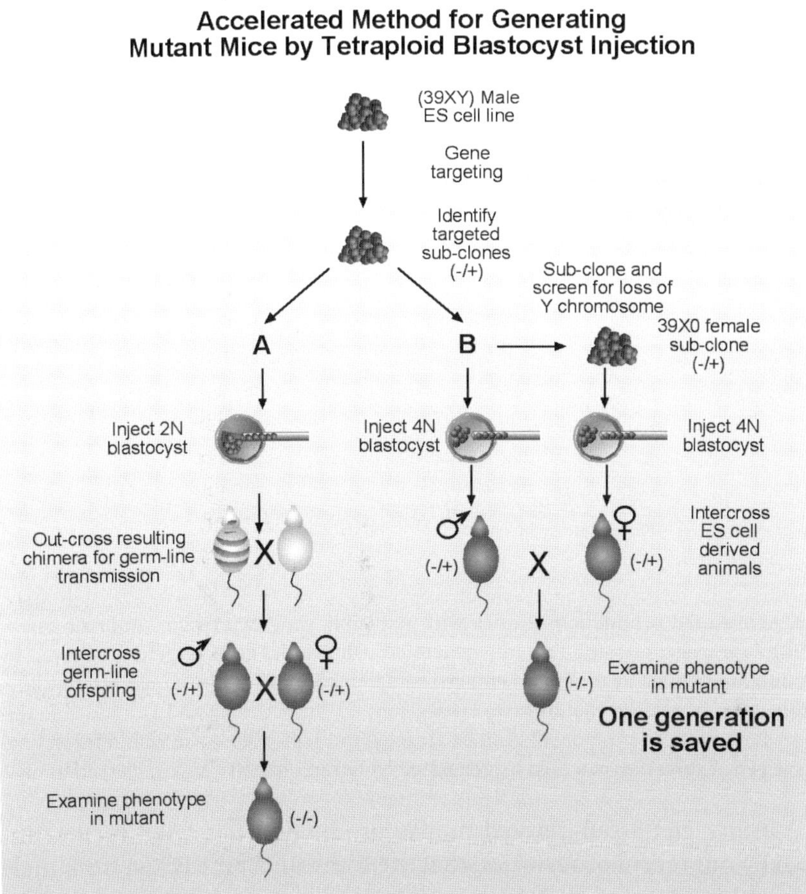

Fig. 3.2. A,B Standard and accelerated production of mutant mice from ES cells. **A** Standard production of mutant mice from heterozygous ES cells requires the generation of chimeric founder animals by introducing targeted male ES cells into diploid blastocysts. Chimeric founders must then be out-crossed to fix the mutation in the male and female germline. These male and female heterozygous offspring are finally intercrossed to produce homozygous mutant progeny. **B** The accelerated production of homozygous mutant offspring can be accomplished by isolating 39X0 sub-clones from targeted 40XY ES cell lines and subsequent production of ES-cell-derived males and females by tetraploid embryo complementation. These heterozygous mice can then be immediately intercrossed to produce homozygous mutant offspring, considerably shortening the time required to generate experimental animals

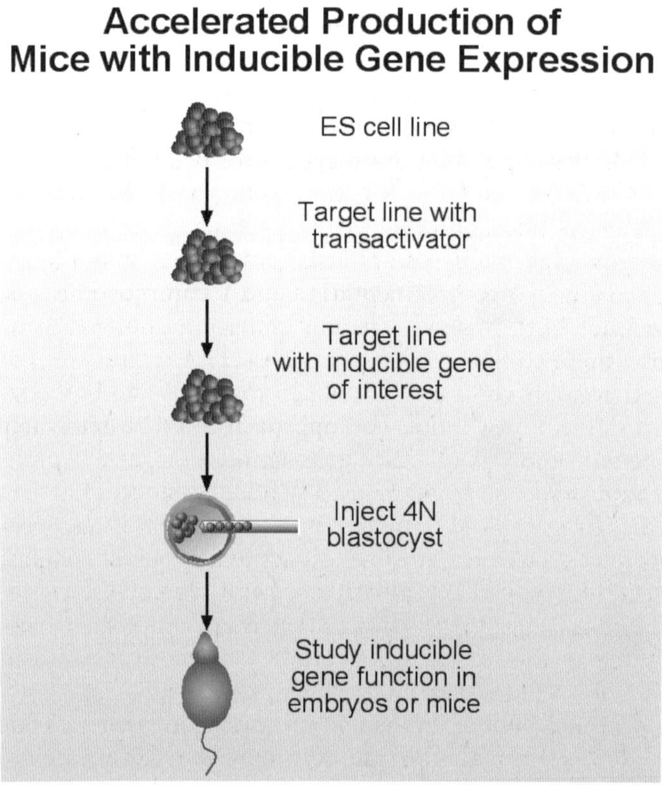

Fig. 3.3. Accelerated production of mice with inducible gene expression. Animals carrying multiple transgenes can be rapidly generated using serial gene targeting followed by the production of mice by tetraploid embryo complementation. For instance, tetracycline-inducible gene expression requires two transgenes, a transactivator and an inducible target gene. Both of these components can be targeted to F1 ES cells. Mice with tetracycline-inducible gene expression can then be generated by tetraploid embryo complementation

The utility of ES cell-derived mouse production has been further increased by our recent observation that the Y chromosome is lost from male ES cells at high frequency (Eggan et al. 2002). Taking advantage of Y chromosome loss allows 39X0 female mice to be produced from targeted male ES cell lines. These heterozygous X0 females are fertile when mated with males derived from 40XY ES cell lines carrying the same mutation (Eggan et al. 2002; Cattanach 1962). This strategy allows homozygous mutant mice to be produced in a single breeding cycle, thereby greatly expediting production of homozygous mutant offspring (Fig. 3.2B).

Importantly, 39X0 subclones can be readily identified from male ES cell lines using established cell culture and molecular techniques, allowing the strategy described above to be utilized by any laboratory already carrying out gene-targeting experiments (Fig. 3.2B) (Eggan et al. 2002). We find that the additional work required for identification of X0 sub-clones is largely compensated by the reduction in breeding cycles needed to produce mice that are homozygous for a mutation or that carry multiple transgenes.

The conventional derivation of mice homozygous for two mutant alleles of interest involves the independent production of targeted ES cell lines, the generation of chimeric mice, followed by two cycles of mating to yield compound heterozygous and finally compound homozygous mutant animals. Using tetraploid embryo-complementation and Y chromosome loss, it is possible to produce both male and female compound heterozygous mutant mice. Mating these compound heterozygotes allows production of mutant mice with all possible combinations of heterozygous and homozygous genotypes in a single mouse cross. This approach could be extremely beneficial for the genetic analysis of entire gene families or genetic pathways, allowing the rapid exploration of many potential phenotypes.

Production of mice by tetraploid embryo complementation is also useful for the rapid production of mice in which bipartite transgenic systems, such as those used for the inducible regulation of gene expression, are required. After two rounds of gene targeting, embryos and mice can be directly produced, allowing the temporal regulation of gene expression in embryos and adults (Fig. 3.3) (Seibler et al. 2003). Another exciting possibility is the direct production of embryos and mice carrying genetic constructs that allow the conditional expression of small interfering RNAs. This approach has recently been demonstrated to enable the functional knockdown of gene expression in ES-derived animals, allowing loss-of-function type analysis to be performed without the time-consuming mating strategies normally required for making mutant mice (Ventura et al. 2004).

In addition to accelerating genetic analysis in mice, the production of female mice from male ES cell lines via tetraploid blastocyst injection (Fig. 3.2B) allows female germ-line transmission of targeted mutations or transgenes that inhibit spermatogenesis (Al-Shawi et al. 1991; Wang et al. 2001). This method also allows female transmission of mutant alleles that cause developmental failure after paternal inheritance, such as paternally expressed imprinted genes (Marahrens et al. 1997).

A critical requirement for the routine use of tetraploid embryo complementation for the production of embryos and mice is that ES cells retain their developmental potency after extensive genetic manipulation. We have previously demonstrated that F1 ES cell lines that were subjected to one or two rounds of selection produced ES cell-tetraploid mice as efficiently, or even more efficiently, than the parental cell line (Eggan et al. 2001, 2002). Convincingly, even after five consecutive rounds of gene targeting, the F1 ES cells were able to generate ES cell-tetraploid mice, establishing the efficacy of the strategies we have outlined here (Eggan et al. 2002).

Before describing the methods use to execute the strategies described above, it is important to consider that the offspring of F1 ES cell-tetraploid mutant mice are a genetically heterogeneous population of F2 animals. If mutant animals with an inbred genetic background are required for quantitative or qualitative analysis of a particular phenotype, substantial backcrossing would be necessary, abrogating the benefits of this approach.

A complete flow chart and task manager for the production of embryos and mice by tetraploid embryo complementation can be seen in Fig. 3.4. As most investigators will undoubtedly begin their experiments with the culture and genetic manipulation of ES cells before proceeding to work with embryos and animals, we will begin with a brief description of the methods that we routinely use for ES cell culture. Proper ES cell culture is critical as poor culture techniques or careless handling of ES cells can cause them to differentiate, and prohibit them from giving rise to ES-cell-derived mice.

3.2 ES Cell Culture

3.2.1 Derivation, Culture and in vitro Gene Targeting of F1 ES Cells

We derive, culture and perform gene targeting in F1 ES cells using standard methods. ES cells are cultured in DMEM (Dulbecco's Modified Eagle's medium; Gibco, Rockville, MD) with 15% fetal calf serum (FCA; Hyclone. www.hyclone.com), 0.1 mM non-essential amino acids (Gibco), 2 mM L-glutamine, 50 IU Penicillin, 50 IU Streptomycin (Gibco) and 0.1 mM β-mercaptoethanol (Sigma, St. Louis, MO), 1,000 U/ml leukemia inhibitory factor (LIF; Chemicon, www.chemicon.com) on gelatinized tissue-culture ware (Falcon, www.bdbiosciences.com). We routinely use gamma-irradiated primary mouse embryo fibroblasts (MEFs) as feeders rather then transformed cell lines. We also often include the MEK kinase inhibitor PD98059 (Cell Signaling Technology. www.cellsignal.com) in our culture media at a final concentration of 50 μM, which seems to aid maintenance of ES cell pluripotency. A complete description of ES cell culture methods is beyond the scope of this chapter but can be found elsewhere (Hogan et al. 1994; Nagy et al. 2002).

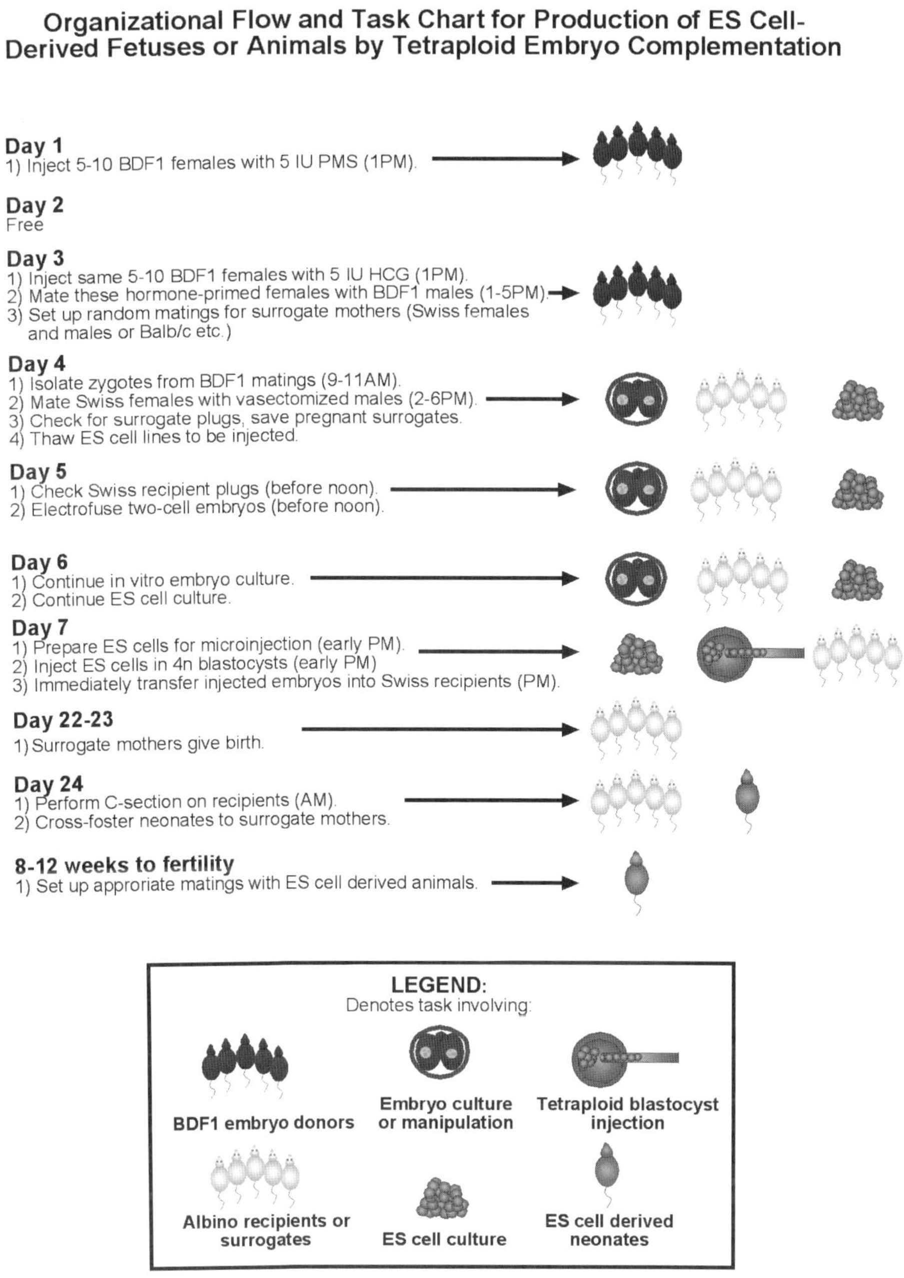

Fig. 3.4. Flow and task chart for production of mice and embryos by tetraploid blastocyst injection. Note that the time in days listed refers to the day on which the various steps in the procedure should be carried out relative to the entire procedure. For instance, Cesarean (C)-section of recipient mothers should be performed on days post coitum (dpc) 19.5 of their pregnancy but day 24 of the overall procedure

3.2.2 Sub-cloning of ES Cells to Identify 39X0 Derivatives of Targeted Cell Lines

In efforts to accelerate the production of mutant mice it can be advantageous to isolate 39X0 sub-clones of a targeted male cell line. Isolation of these sub-clones allows direct production of mutant female mice for breeding (Fig. 3.2). To isolate 39X0 derivatives of targeted ES cell lines, we perform simple dilution sub-cloning. To this end, we expand the 40XY targeted sub-clone, trypsinize, count the cells and plate 5,000 ES cells in a 10 cm dish preplated with irradiated MEFs. Between 5 and 8 days after plating, well-defined ES colonies can usually be seen. We then pick 200–300 colonies, and expand them on female MEFs for cell freezing and DNA isolation. Picking 200–300 clones is usually sufficient to identify several X0 sub-clones, which we have found arise at a frequency of 2%. It is essential to expand the ES cell sub-clones on female MEFs because contaminating Y chromosome DNA from male MEFs may confound Y chromosome genotyping. To identify the 39X0 sub-clones, we perform Y chromosome genotyping by Southern hybridization on crudely prepared genomic DNA (Laird et al. 1991) although PCR is also effective. After identification, frozen sub-clones can be thawed and expanded. Female mutant mice can then be produced by tetraploid embryo complementation from these cells.

3.2.3 Y Chromosome Genotyping by Southern Hybridization

For Y chromosome Southern analysis of ES cell subclones, genomic DNA is digested with EcoRI, blotted to a membrane and probed with a 720 bp MboI fragment of the plasmid pY2 (Lamar and Palmer 1984). Alternatively, DNA can be digested with PstI, blotted and probed with the 1.5 kB EcoRI fragment of the plasmid pY353 (Bishop and Hatat 1987). Both of these probes hybridize to high-copy repetitive sequences located primarily on the Y chromosome, giving a bright signal at a range of molecular weights (Fig. 3.5). It is advisable to always run a female DNA control on each blot so that the male specific pattern of bands can be easily determined (Fig. 3.5A). 39X0 sub-clones are then recognized easily as they lack these male-specific hybridization signals but retain some cross-reactivity with a small number of autosomal repeats. Photographing the agarose gel before blotting can also help identify X0 clones, preventing confusion with under-loaded lanes (Fig. 3.5B,C).

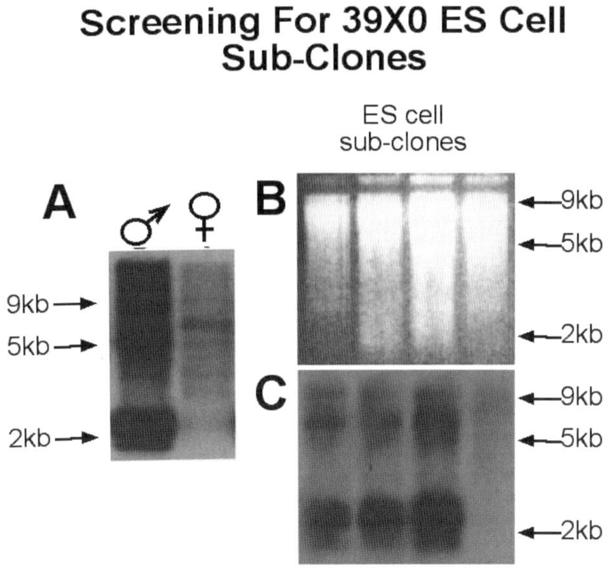

Fig. 3.5. A–C Screening ES cell sub-clones for loss of the Y chromosome by Southern hybridization. Using Southern analysis with a probe specific to repetitive elements on the Y chromosome, male and female cell-lines can be easily distinguished (**A**). By comparing the DNA loading on the gel (**B**) with the results of Southern hybridization with the repeat probe (**C**), this genotyping method can be easily adopted to discern between 39XY ES cell sub-clones (first three lanes) and a 39X0 sub-clone (last lane). These 39X0 sub-clones can then be used to produce female mice (see Fig. 3.2B). (Adapted from Eggan et al. 2001)

3.2.4 Identification of 39X0 Subclones by PCR

PCR genotyping for the Y chromosome can also be carried out rapidly on crude genomic DNA with primers that amplify sequences from the Y-linked Zfy gene (Zfy primer 1: 5'-GATAAGCTTACATAATCACATGGA, Zfy primer 2: 5'-CCTATGAAATCCTTTGCTGC). Primers specific for the Y chromosome EST Tet35 have also been used successfully to screen for loss of the Y chromosome (A. Bortvin, personal communication). The following primers are used for TET35 PCR, primer 1: 5'-CTCATGTAGACCAAGATGACC, primer 2: 5'-GGAATGAATGTGTTCCATGTCG. For these primer sets, we perform PCR for 30–35 cycles, annealing at 60 °C for 10–15 s and extending for 30 s at 72 °C. PCR products are then resolved on 1% agarose gels to assay for the presence of the Y chromosome.

3.2.5 Karyotyping of ES Cell Lines during Serial Gene Targeting

We and others have observed that certain karyotypic abnormalities in ES cells may interfere with both production of ES cell-tetraploid mice and germ-line transmission of targeted mutations (Nagy et al. 1993; Wang et al. 1997; Marahrens et al. 1997). Surprisingly, we have found that the majority of chromosomal abnormalities present in most, if not all, ES cell lines, did not interfere with their potency to generate adult mice by tetraploid embryo complementation (Eggan et al. 2002). Our chromosomal analyses have revealed that the abnormalities found in donor ES cell lines are not present in the animals derived from them. It may be that only subsets of the injected cells – those with a normal karyotype – contribute to embryo formation. In contrast, cells with some karyotypic abnormalities may be selected against during development or subsequent ES cell culture. However, any chromosomal abnormality resulting in a growth advantage in vitro can be expected to completely overgrow the population of karyotypically normal cells and eliminate a cell line's competence to generate mice (Liu et al. 1997). Therefore, when carrying out serial rounds of gene targeting, we often perform chromosomal analysis to exclude lines dominated by cells with abnormal karyotypes. If these precautions are taken, it may be that there is no limit to the number of times an F1 ES cell line can be genetically manipulated before mutant mouse production.

3.3 Production of Tetraploid Embryos

A number of methods have been used successfully to produce tetraploid embryos from diploid embryos, including preventing cytokinesis with Cytochalasin B (Edwards 1958) and by chemically inducing cell-cell fusion with polyethylene glycol (Eglitis 1980). However, embryos for tetraploid embryo complementation are generally created by electrically fusing the two blastomeres of a two-cell embryo. The popularity of this method can be ascribed to the simplicity by which a large number of tetraploid embryos can be made, and the high efficiency with which these embryos develop to the blastocyst stage. To summarize the protocol, diploid zygotes are isolated from super-ovulated females, cultured to the two-cell stage, when electrofusion is performed, and then further cultured to the blastocyst stage, at which time ES cells are injected (Figs. 3.1, 3.4).

In our experience, one of the most critical determinants in successfully producing mice by tetraploid embryo complementation is proper in vitro culture and maintenance of the preimplantion embryos. If special attention is not paid to in vitro culture technique, embryo viability may be compromised, making the generation of tetraploid blastocysts both inefficient and frustrating. Development of tetraploid embryos to the blastocyst stage should be at least 80–95% efficient. If this efficiency is not routinely achieved, post-implantation viability may be compromised and mouse production will be extremely difficult. To prevent day-to-day variability, and to ensure consistency in embryo culture we now use the M2 and KSOM media supplied by Specialty Media (Phillipsburg, NJ). MCZB medium (see section on Methods, Equipment and Techniques in Chap. 4 of this volume, Cloning the Laboratory Mouse by Nuclear Transfer) can also be used routinely for embryo culture; however, it must be freshly prepared at least every 2 weeks.

3.3.1 Isolation and In Vitro Culture of Preimplantation Embryos

We isolate fertilized zygotes from B6D2F1 female animals because they provide a large number of embryos after superovulation, and because these embryos are extremely tolerant to in vitro culture (Fig. 3.4). Females of 8–12 weeks old (NCI/Charles River, www.criver.com) are maintained in our colony for at least 2 weeks before superovulation. This delay ensures time for the females to undergo day-night cycle adjustment. To induce superovulation, we first inject 5 IU pregnant mare's serum (PMS; Calbiochem, San Diego, CA) [intraperitoneal (IP) injection], followed 48 h later by 5 IU human chorionic gonadotropin (HCG; Calbiochem) (Fig. 3.4). These injections are performed in the early afternoon. After HCG injection, females are individually housed/mated with a B6D2F1 male overnight. The next morning, 20–24 h after HCG injection, fertilized zygotes are isolated from the oviducts of plugged females. In general, superovulation of 5–10 BDF1 females is sufficient for a blastocyst injection, allowing isolation of between 150 and 200 fertilized zygotes.

It is strongly recommended that all preimplantation embryo culture be carried out in microdrops of medium on standard bacterial Petri dishes (Falcon) under mineral oil (buy at your local pharmacy, Squib or similar). To prepare micro drop culture dishes, pipette 25–50 μl drops of embryo culture medium onto the dish and then cover the drops with mineral oil. The drops of medium in the Petri dish should be completely covered with mineral oil to prevent evaporation.

For zygote isolation, sacrifice the superovulated females by approved methods and dissect the oviducts into a Petri dish, containing several 30 μl drops of M2 medium (Specialty Media) and several drops of M2 medium supplemented with 0.1% w/v bovine testicular hyaluronidase, covered with mineral oil (Sigma). Viewing the dissected oviduct submerged in the mineral oil under a standard transillumination-dissecting microscope, the ampulae containing the fertilized zygotes should be easily observed. Using a pair of sharp forceps, the ampula can be torn, releasing the zygotes, still surrounded by their cumulus masses, into the drops of M2 media containing hyaluronidase. We find that dissecting under mineral oil allows for rapid quantitative isolation of zygotes with minimal hypertonic shock to the embryos. After several minutes in M2 with hyaluronidase the cumulus masses surrounding the zygotes will dissociate, allowing the zygotes to be isolated.

Following cumulus mass dissociation, the zygotes should be picked up with a pulled glass micro capillary pipette. These pipettes can be made from 50 μl glass capillaries pulled over a Bunsen burner and can be controlled with a simple mouth controlled aspirator tube assembly (VWR, Marietta, GA). After pick up, the zygotes are washed through several drops of M2 medium to both eliminate the hyaluronidase and to remove excess cumulus cells. Zygotes are then immediately transferred to microdrops of KSOM medium (Specialty Media) and placed at 37 °C with an atmosphere of 5% CO_2 in air for overnight culture (Figs. 3.1, 3.4). In general, this entire procedure, from sacrificing animals to placing the embryos into the incubator should take no longer than 20–30 min for practiced hands. If the embryos are exposed to room temperature conditions or hyaluronidase for longer periods of time, their viability may be compromised. If it is difficult to finish the procedure in the suggested time frame, sacrificing animals and isolating the embryos in two groups is the best solution.

3.3.2 Electrofusion of Two-cell Embryos

The most popular and effective method for production of tetraploid embryos is electrofusion of the two-cell embryo blastomeres. Application of a direct current (DC) electrical-pulse across a lipid bi-layer leads to the opening of small holes in the membrane. Because of the direct juxtaposition and tight junctions between the membranes of the two cells, these openings often resolve into small cytoplasmic bridges. Since a spherical shape is the lowest energy equilibrium-state for the membrane, these bridges continue to widen until the two cells have completely fused to form one continuous cytoplasm containing two nuclei (Fig. 3.6A–H). At the next mitosis, nuclear envelope breakdown of the two diploid nuclei occurs and all of the chromosomes align on a single spindle. Following cytokinesis these chromosomes coalesce into a single tetraploid nucleus. Thus two diploid-cells become one tetraploid-cell (see Fig. 3.1).

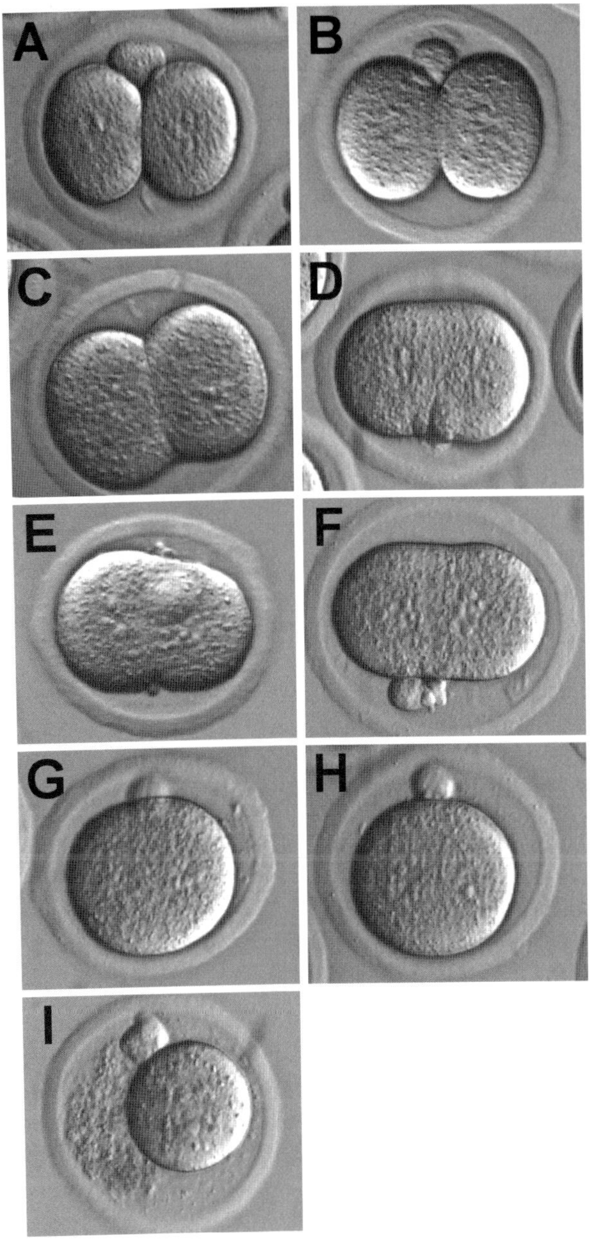

Fig. 3.6. A–I Electrofusion of two-cell embryos. **A–H** Two-cell embryos 10 min after direct current (DC) fusion at various stages of cell fusion. **I** A two-cell embryo in which one of the two blastomeres has died during the electrofusion procedure. These diploid embryos must be removed to ensure that they do not confound the results of the tetraploid embryo complementation experiment (see text)

For efficient cell-fusion, the electrical field must be perpendicular to the membrane between the two cells of the embryo. The two-cell embryo can be aligned in this manner either manually via micromanipulation or electrically with the application of an alternating current (AC, see below). Devices capable of applying sufficient AC and DC electrical fields are available from several sources [BLS, Hungary (www.bls-ltd.com); BTX (www.btxonline.com)].

Under the conditions described below it will be apparent within several minutes whether membrane fusion has begun, although it may take as long as an hour for all of the embryos to complete fusion. Using these methods we routinely observe fusion rates of 95% without embryo lysis. However, embryos in which one of the two blastomeres has lysed due to the electrical-treatment are occasionally observed. Lysis of one of the blastomeres results in a one-cell diploid embryo that can be easily confused with the tetraploid embryos (Fig. 3.6I). Unfortunately, these diploid one-cell embryos often continue to develop into diploid blastocysts and can confound interpretation of tetraploid embryo-complementation experiments if not eliminated. It is, therefore, critical to remove these embryos, as well as un-fused two cell embryos, in the hour following fusion.

The precise field strengths and electrofusion conditions required for efficient cell-fusion with minimal damage to the embryo may vary between pulse generators, and should be determined experimentally. A reasonable rule of thumb is that if the rate of cell-fusion is low, then the DC voltage or pulse width should be increased. In contrast, if embryo lysis or poor preimplantation development occur, voltage and/or pulse width should be decreased.

3.3.3 Electrofusion by AC Alignment and DC Pulse

Alignment of the embryos with alternating current (AC), followed by direct current (DC) fusion is a convenient method to quickly produce a large number of tetraploid embryos. When acting on a two-cell embryo, the AC field generates a membrane potential, polarizing both cells within the embryo. This polarization has two important effects. Firstly, it causes adjacent membranes to become attracted to one another, forming a tighter junction between the cells; this tighter cell-cell contact helps mediate DC cell-fusion. Secondly, polarization causes the embryo to begin to rotate until the cleavage plane of the two blastomeres is perpendicular to the electrical field and to the subsequent DC pulse. Using this method, 50–100 two-cell embryos can be simultaneously fused with only 5–10 min of work.

In order to generate an AC field in the medium, a non-electrolyte salt (generally mannitol) is required. For DC fusion with AC alignment we use a fusion medium composed of 0.3 M mannitol, 0.1 mM MgSO4, 50 μm $CaCl_2$ and 3% bovine serum albumin at pH 7.4. The medium should be stored at −20 °C and is generally good for several months. This medium should be thawed the day of the experiment and care should be taken to ensure the mannitol, which often precipitates, is fully dissolved. Before electrofusion, the embryos are first equilibrated into fusion medium. This is accomplished simply by placing the embryos at the top of a large drop of fusion medium (1 ml) and allowing them to settle to the bottom. After equilibration, the embryos may be transferred into the fusion chamber containing the mannitol medium.

For the AC/DC fusion method, we use a fusion chamber consisting of two parallel platinum wires (1 mm gap width) immobilized on a glass Petri dish (such as GSS-1000, BLS Ltd. or BTX microslide 450–1). The chamber is first flooded with electrofusion medium; the equilibrated embryos are then placed midway between the two electrodes. The AC field is initiated before the embryos settle to the bottom of the chamber, facilitating rotation of the embryos in the electrical field. We recommend trying an initial AC field strength of 2 AC volts; however, the field strength required to rotate the embryos may vary from chamber to chamber and is greatly influenced by electrode gap distance. The minimal field strength required to rotate the embryos should be identified experimentally and subsequently used. Some embryos may not align in the field due to adherence to the chamber or to one another. These embryos can be nudged with the end of a capillary, allowing them to align. A small percentage of the embryos may never align due to asymmetry between the two blastomeres; such embryos will not undergo cell fusion.

Once the embryos have aligned, the DC pulse (1.5 kV/cm) is applied to initiate cell-fusion. Following the DC pulse, the embryos should be removed from the chamber and washed through several large drops (200 μl) of KSOM and returned to the incubator for long-term culture. After electrical treatment, inspect the embryos every 15 min to identify and pool newly formed tetraploid embryos that have undergone cell-fusion.

3.3.4 Manual Alignment and DC Fusion of Two-cell Embryos

As an alternative to AC alignment, we also suggest using micromanipulators/electrodes to manually align two-cell embryos for fusion. Manual alignment and DC fusion are carried out on an inverted microscope using the lid of a Petri dish as a micromanipulation chamber. Platinum wires

(200–300 μm in diameter) are used as both electrodes and micromanipulators to move the two-cell embryos into position for fusion. A group of 15–25 two-cell embryos is placed on the stage in a 500 μl drop of M2 medium (Sigma). Note that, in this case, the M2 medium is not covered in mineral oil, as mineral oil tends to coat the electrodes. This coating can cause substantial variability in electrical field strength and cell-fusion. Because medium not covered in mineral oil tends to evaporate quickly, changing its isotonicity, it is critical to work rapidly. In order to prevent hypertonic shock, operation on a single group should not take more than 5–10 min. For cell-fusion, embryos are aligned with the interface between their two blastomeres perpendicular to the electrical field, and a single electrical pulse (1.5 kV/cm for 90 μs) is applied to each in turn. After electrofusion, embryos are washed several times through KSOM medium and then cultured in KSOM medium at 37 °C. Each subsequent round of electrofusion should be carried out in a fresh drop of M2 medium. After finishing, remember to check the embryos periodically to identify those that have undergone successful fusion.

3.3.5 Culture of Tetraploid Embryos to Blastocyst Stage

After electrofusion, the embryos should be cultured to the blastocyst stage in KSOM medium. Tetraploid embryos continue to develop with timing similar to their diploid two-cell counterparts. Approximately 24 h after fusion, four-cell tetraploid embryos will undergo compaction (similar timing to compaction of diploid eight-cell embryos). At 48–56 h after fusion, tetraploid embryos should start to develop a blastocoel cavity. Tetraploid embryos should therefore be ready for injection in the afternoon of day 7 of the protocol, or a little more than 72 h after fertilization (Fig. 3.4). Substantial delays in either compaction or blastocyst development may indicate that the cell-fusion parameters were too harsh or that the embryo culture medium has become too old. We generally use KSOM medium for only 7–12 days before preparing a new batch. Media quality seems to be one of the most important parameters for efficient preimplantation development of tetraploid embryos.

3.3.6 Microscope Set-up for Microinjection

We perform blastocyst injections on an inverted microscope with Hoffman modulation contrast optics (Fig. 3.7) using Piezo-assisted microinjection [Primetech Pmm, Ibaraki, Japan, distributed in the United States by Ep-

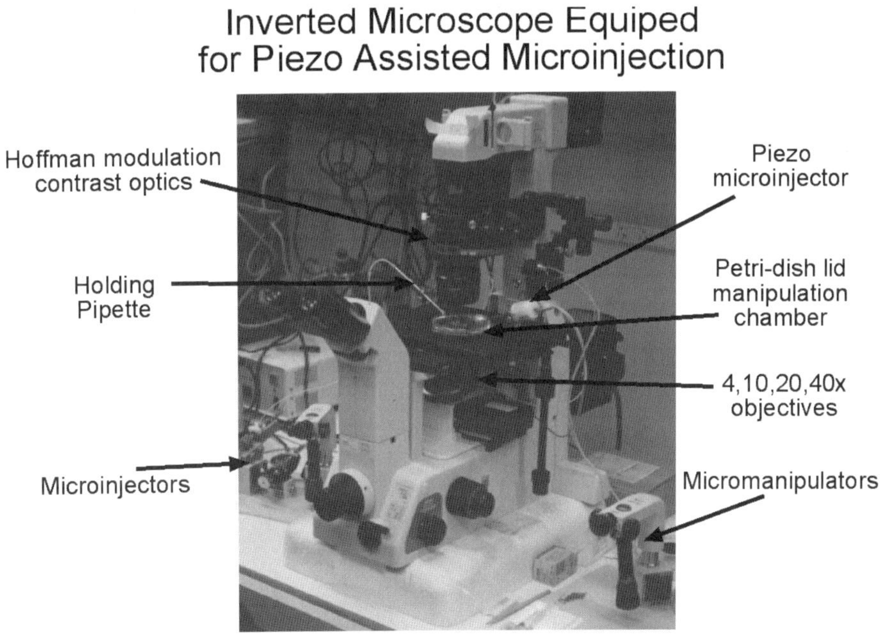

Fig. 3.7. Micromanipulators and inverted microscope set for Piezo micromanipulation. Inverted microscope with Hoffman modulation contrast set for nuclear transfer using the lid of a Petri dish as a micromanipulation chamber. Notice the correct angle of inclination for the Piezo micromanipulator, which is between 15–20 degrees. Also note that all micromanipulation is performed under mineral oil to prevent medium evaporation and hypertonic shock to the oocytes

pendorf (www.eppendorfna.com)]. Although tetraploid blastocysts can be injected using standard techniques and a sharpened injection pipette, we find that the Piezo element (Primetech) allows us to inject a larger number of embryos in a shorter time. In addition, generating flat-tip Piezo injection pipettes is much simpler than making sharpened needles. The disadvantages of using the Piezo element are the expense of buying this apparatus and the added danger of working with small amounts of mercury on a regular basis. As will be explained below, mercury is required to increase the needle mass, allowing it to break into the blastocoel cavity.

The microscope should be equipped with micromanipulators (Narishige NT 88ne, or similar; www.narishige.co.jp) and microinjectors. We use an IM6–2 microinjector (Narishige) for Piezo-assisted microinjection because it can withstand the rather large backpressure generated by the mercury. There is no specific requirement for suction control on the holding pipette. We use an IM9-B microinjector (Narishige). The lines of these two microinjectors can be filled with water, which we find more convenient than oil.

3.3.7
Preparation of Microinjection Instruments for Piezo Microinjection

For ES cell injection into tetraploid blastocysts, we use a flat-tip microinjection pipette with an internal diameter of 15–20 μm. The blastocyst to be injected is held with a standard holding pipette. We find that these pipettes are very easy to make if you have the correct equipment; however, they can also be purchased from Humagen (Charlottesville, VA). We prepare microinjection instruments using a Flaming-Brown micropipette puller with a box filament [model P-87, Sutter (www.sutter.com)] and Microforge de Fonbrune model M-830 (Narishige). Before preparing instruments, a small glass bead (75–100 μm) should be applied to the microforge by melting a small piece of glass microcapillary to the heated filament (Fig. 3.8B).

The first step in preparing the instruments is to pull a 0.75 mm inner diameter microcapillary (B100–75–100; Sutter) in the micropipette puller (Fig. 3.8A). After the micropipette is pulled, it is placed on the microforge with a region of the pipette of the appropriate internal diameter (15–20 μm) just touching the top of the glass bead (Fig. 3.8B). The heating element of the microforge is then briefly turned on and then off. This should cause the needle to melt onto the bead, and then, as the filament cools, cleanly break the needle (Fig. 3.8C). If the capillary does not attach to the glass bead then the heat should be increased. Conversely, if the break point of the needle becomes distorted, or is not clean or flat, the heat should be decreased. After breaking the needle, the excess glass is brushed away with a small artist's brush or pair of forceps (Fig. 3.8D). In order to bend the pipette, the tip is advanced well past the heating filament, without touching it (Fig. 3.8E), and heat is then applied (Fig. 3.8F). We generally bend the pipette 10–20 degrees. It is critical that the internal diameter of the injection pipette at the tip is sufficiently large that the cells are not constricted inside the injection needle (15–20 μm). If the cells have to be squeezed at the tip of the pipette as they are being picked up, they can become lysed by the vibrating action of the Piezo element during blastocyst injection.

For the action of the Piezo element to be effective, the microinjection pipette must have a high mass. For instance, if the Piezo element is applied to an empty pipette, the pipette will simply bounce off of the blastocyst. To provide the needle with enough inertia to penetrate the blastocoel cavity during microinjection, the injection pipette is back-filled with 3–5 μl

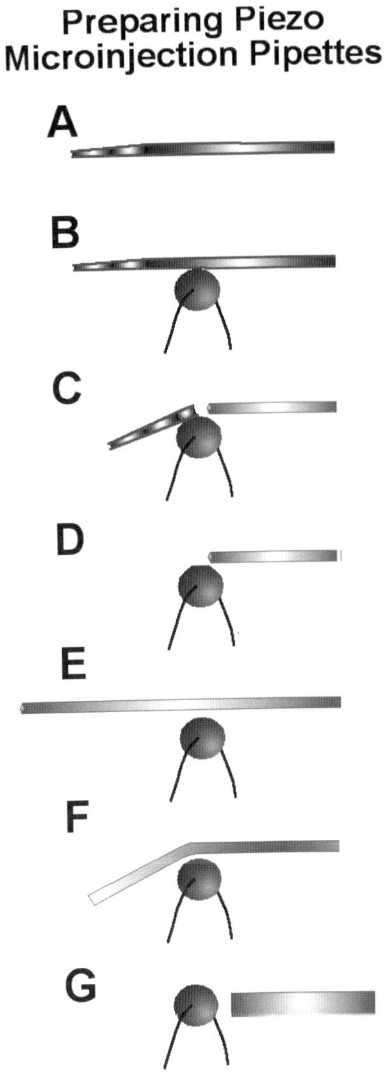

Fig. 3.8. A–G Preparation of Piezo microinjection pipettes. After the micropipette is pulled (**A**), it is placed on the microforge such that the region of the pipette with the appropriated internal diameter (15–20 μm) is just touching the top of the glass bead (**B**). The heating element of the microforge is then briefly turned on and then turned off. This should cause the needle to melt onto the bead and then in turn cleanly break the needle (**C**). Using a pair of forceps, the excess glass should then be brushed away (**D**). After the needle is broken appropriately, it should be bent. To bend the needle, first advance the tip of the needle well past the heating filament without touching the filament (**E**), and then apply heat (**F**). To fire polish a holding pipette (**G**), it should be held in the position shown with heat applied to the microforge filament

mercury (www.jtbaker.com). Mercury is hazardous! Great care should be taken any time mercury is being handled and appropriate protocols for spill clean up should be in place before work begins. To backfill the injection pipette, pipette 3–5 μl mercury into a 1 ml syringe with a small piece of silicon tubing attached to the tip. The open end of the silicon tubing is then slipped onto the back of the injection pipette and the mercury can be loaded. Great care should be taken while loading the pipette, as it is easy to spill mercury during this procedure. After loading the pipette, attach it to the micromanipulator as for normal blastocyst injections. For blastocyst injection of ES cells, the Piezo element is set at an intensity of 4–7 with a speed of 2–5.

A holding pipette for immobilizing the blastocyst during injection is prepared by a similar method, but the internal diameter at the break should be 40–60 μm. Before bending, the holding pipette should be fire polished by slowly moving the broken end of the pipette towards the heated filament of the microforge, head-on, as depicted in Fig. 3.8G. This will cause the end of the pipette to become rounded and decrease the internal diameter of the opening to around 15 μm (see Fig. 3.9B). After polishing the pipette, it should be bent to an angle of around 20 degrees.

3.3.8 Piezo-Micromanipulator Injection of Tetraploid Blastocysts with ES Cells

For blastocyst microinjection, we use the lid of a standard bacterial Petri dish flooded with mineral oil as a microinjection chamber (see Fig. 3.7 and also Chap. 4 Cloning the Laboratory Mouse by Nuclear Transfer, Fig. 4.2). The Petri dish lid contains several drops of microinjection (ES cell medium, omitting LIF) and one drop of the same medium supplemented with 10% (w/v) polyvinylpyrrolidone (PVP; MW 360,000; ICN Biochemicals, Aurora, OH). The medium with PVP should be prepared well in advance as it takes some time for the PVP to go into solution. Once tetraploid morulae have cavitated to form blastocoel cavities (cavitated), 10–15 blastocysts are placed in a drop of injection medium. The ES cells, which have been trypsinized and held on ice, are then diluted into a separate drop of medium.

The microinjection pipette is lowered down into the drop of medium containing PVP and any remaining air is expelled from the tip of the needle. The expulsion of several drops of mercury into the PVP is also routine (Fig. 3.9A). The inside and outside of the injection pipette should then be coated with PVP by aspirating the medium into the pipette and then blowing out a few small drops of mercury. This action should be repeated sev-

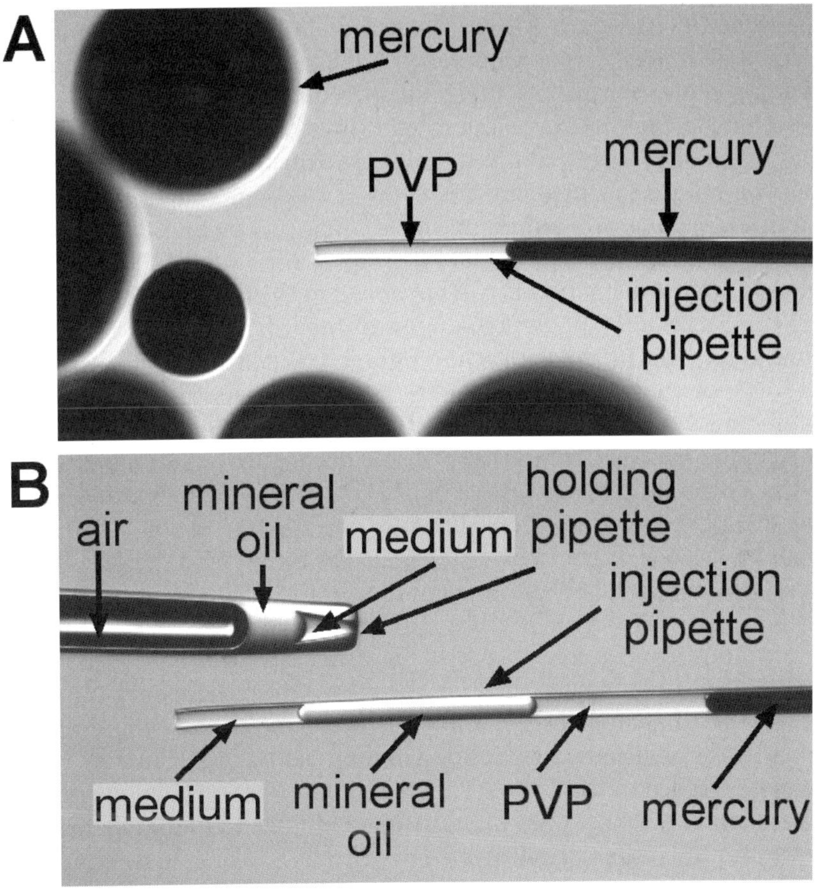

Fig. 3.9. A,B Set up of instruments on the microscope for Piezo microinjection. **A** The microinjection needle, which has been carefully backfilled with mercury, is first lowered into a drop of PVP in the micromanipulation dish. A few drops of mercury are then expelled into the drop of PVP and the pipette thoroughly coated inside and out with PVP. **B** After coating the needle with PVP, a small cushion of mineral oil is pulled into the pipette and the pipette is then moved into a drop of injection medium containing embryos or ES cells. The holding pipette is then lowered into the medium. In general, we fill the holding pipette with air and a small cushion of mineral oil to help control suction on the holding pipette

eral times (Fig. 3.9A). After washing the pipette, a small amount of PVP is picked up, followed by a small drop of mineral oil from the dish and the pipette is then moved into a drop of injection medium. This will result in an injection pipette similar to that shown in the bottom of Fig. 3.9B. After the injection pipette has been brought into the injection medium, the holding pipette can be lowered into the same drop. It is advisable to avoid moving the holding pipette into the drop of medium containing PVP. A large amount of PVP will adhere to the holding pipette and this PVP can be toxic to the embryos.

After set up, the instruments can be moved to the drop of medium containing the ES cells. By applying suction to the microinjector, 50–100 ES cells are picked up in the end of the injection pipette (Fig. 3.10A). This allows 5–10 blastocysts to be injected at the same time.

Following ES cell pick-up, the stage is moved until the drop of medium containing the embryos is in view. A single blastocyst can then be immobilized with the holding pipette (Fig. 3.10B). The injection pipette containing the ES cells is then pressed against the zona pellucida and a brief pulse of the Piezo element is applied. While the Piezo is being applied, the injection needle is pushed through the zona and trophectoderm into the blastocoel cavity (Fig. 3.10C). About 10–12 ES cells are expelled from the injection pipette and pushed against the inner surface of the blastocyst (Fig. 3.10D). The pipette can then be withdrawn from the embryo and the next blastocyst injected (Fig. 3.10E). If at any time the injection pipette becomes sticky or clogged, the holding pipette can be raised and the injection pipette washed in the drop of PVP medium. After washing, the instruments are returned to the drop of injection medium and more cells may be picked up for injection.

Following injection of an entire group of blastocysts, the embryos are returned to KSOM medium and placed at 37 °C until transfer to recipient females. In order to produce mutant mice routinely, one should expect to produce and inject at least 50 and preferably 75–100 tetraploid blastocysts per experiment.

3.3.9 Embryo Transfer to Recipient Females

After blastocyst injection, 10–12 tetraploid-ES cell embryos are transferred to each uterine horn of 2.5 day post coitum (dpc) pseudopregnant Swiss Webster recipients (see (Hogan et al. 1994) for a complete description of embryo transfer). Pseudopregnant recipient mice are generated by mating Swiss females to vasectomized males. Vasectomized males can be purchased from a vendor such as Taconic (www.taconic.com) and should be

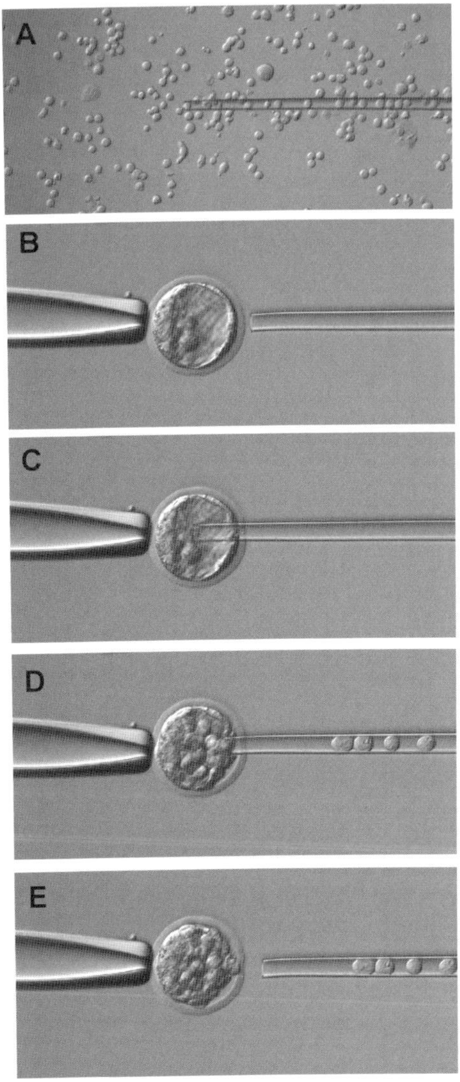

Fig. 3.10. A–E Piezo micromanipulator injection of ES cells into a blastocyst. **A** ES cells are collected in a flat-tipped microinjection pipette that is slightly larger then the ES cells. If the diameter of the needle is not larger then the ES cells, the ES cells may lyse during microinjection. After ES cell pick-up, the instruments are moved to a drop of medium containing several blastocysts. **B** The blastocysts to be injected should be held adjacent to the inner cell mass (ICM) if present. However, the ICM will often be absent in tetraploid blastocysts. **C** The flat-tipped injection pipette is pressed against the zona and the Piezo micromanipulator is applied, allowing the pipette to enter the blastocoel cavity. After ES cell deposition (**D**), the needle is removed (**E**) and the next blastocyst can be injected

tested for sterility before general use. We find 30–35 g to be an ideal weight for the recipient females. See Fig. 3.4 for details on the appropriate timing, relative to other aspects of tetraploid embryo complementation, for producing pseudopregnant recipients.

3.3.10 Cesarian Section and Cross Fostering of ES Cell Tetraploid Mice

Although it is by no means a requirement, we find it beneficial to deliver pups by performing a Cesarian-section (C-section) on the embryo transfer recipient on the morning of dpc 19.5. We find an increased incidence of neonatal complications when these mothers give birth to small litters, and the increased level of neonatal care we provide allows a higher proportion of pups to survive. To deliver neonates, the recipient mother is sacrificed, and the pups are rapidly removed from the uterus. Special care should be taken when cutting the umbilical cords of these neonates as the umbilicus is occasionally herniated, which can cause excessive bleeding. After removal from the uterus the pups should be quickly dried with a Q-tip, paying special attention to clear all fluid from the airways. Finally, the pups are placed under a heat lamp until they are cross-fostered.

We cross-foster ES cell-tetraploid neonates delivered by C-section to BALB/c mothers that have given birth either the same or the previous day (see Fig. 3.4 for timing of mating). We have also cross-fostered to CD-1, ICR and Swiss mothers with success. These strains all have dispositions that promote the acceptance of fostered pups, while C57/BL/6 and 129 Sv mothers often reject fostered neonates. To cross-foster, briefly mix the neonates with a small amount of bedding from the foster mother's cage, remove the foster mother's litter and quickly replace it with the ES cell-tetraploid neonates. If the number of ES cell-tetraploid neonates to foster is small, we will generally also return several of the foster mother's pups.

3.4 Final Words

Using the methods described here, 10–25% of injected blastocysts can be expected to survive to term, depending on the ES cell line used (Tables 3.1, 3.2). Importantly, if an ES cell with an F1 genotype is used, around 85% of ES cell-tetraploid neonates can be expected to survive to adulthood (Fig. 3.1B). In contrast, only the rare neonate will survive if cell lines derived from an inbred strain are used (Fig. 3.1A). Thus, if the goal is to produce live mice, it is advisable that F1 cell lines be used. Finally, a question of general concern for the experimental use of ES-cell-derived mice has been whether they have phenotypes that would complicate their use in developmental genetic analysis. Recent experiments suggest that this is not the case: (Schwenk et al. 2003) carried out extensive phenotypic analysis of a large cohort of mice and found that these mice tend to have elevated head-to-rump lengths and a corresponding increase in body weight, but no other overt phenotype (Schwenk et al. 2003). Thus, ES-cell-derived mice provide a stable background in which to observe genetic influences on a complete spectrum of phenotypes.

Acknowledgements. We would like to thank A. Bortvin and C. Beard for providing primer sequences specific to the Tet35 and Zfy genes, respectively, and K. Hochedlinger and W. Rideout for many helpful discussions.

References

Al-Shawi R, Burke J, Wallace H, Jones C, Harrison S, Buxton D, Maley S, Chandly A, Bishop JO (1991) The herpes simplex type 1 thymidine kinase is expressed in the testes of transgenic mice under the control of a cryptic promoter. Mol Cell Biol 11:4207–4216

Bishop CE, Hatat D (1987) Molecular cloning and sequence analysis of a mouse Y chromosome RNA transcript expressed in the testis. Nucleic Acids Res 15:2959–2969

Cattanach B (1962) Xo mice. Genet Res 3:487–490

Edwards RG (1958) Colchicine-induced heteroploidy in the mouse. ii the induction of tetraploidy and heteroploidy. J Exp Zool 137:349–362

Eggan K, Akutsu H, Loring J, Jackson-Grusby L, Klemm M, Rideout WM III, Yanagimachi R, Jaenisch R (2001) Hybrid vigor, fetal overgrowth, and the viability of mice derived by nuclear cloning and tetraploid embryo complementation. Proc Natl Acad Sci USA 98:6209–6214

Eggan K, Rode A, Jentsch I, Samuel C, Hennek T, Tintrup H, Zevnik B, Erwin J, Loring J, Jackson-Grusby L, Speicher MR, Kuehn R, Jaenisch R (2002) Male and female mice derived from the same embryonic stem clone by tetraploid embryo complementation. Nat Biotechnol 20:455–459

Eggan K, Jaenisch R (2003) Differentiation of F1 embryonic stem cells into viable male and female mice by tetraploid embryo complementation. Methods Enzymol 365:25–39

Eglitis MA (1980) Formation of tetraploid mouse blastocysts following blastomere fusion with polyethylene glycol. J Exp Zool 213:309–313

Hogan B, Beddington R, Constantini F, Lacy E (1994) Manipulating the mouse embryo. 2 edn Cold Spring Harbor Laboratory Press, Cold Spring Harbor, NY

Laird PW, Zidjgerveld A, Linders K, Rudnicki MA, Jaenisch R, Berns A (1991) A simplified mammalian DNA isolation procedure. Nucleic Acids Res 19:4293

Lamar EE, Palmer E (1984) Y-encoded, species-specific DNA in mice: evidence that the Y chromosome exists in two polymorphic forms in inbred strains. Cell 37:171–177

Liu X, Wu H, Loring J, Hormuzdi S, Disreche CM, Bornstein P, Jaenisch R (1997) Trisomy eight in ES cells is a common potential problem in gene targeting and interferes with germ line transmission. Dev Dyn 209:85–91

Marahrens Y, Panning P, Dausman J, Strauss W, Jaenisch R (1997) Xist-deficient mice are defective in dosage compensation but not in spermatogenesis. Genes Dev 11:156–166

Mclaughlin KJ (1993) Production of tetraploid embryos by electrofusion. Methods Enzymol 225:919–930

Nagy, A., Gersenstein, M., Vintersten, K., Behringer, R. (eds) (2002) Manipulating the mouse embryo. 3 edn Cold Spring Harbor Laboratory Press, Cold Spring Harbor, NY

Nagy A, Gocza E, Diaz EM, Prideaux VR, Ivanyi E, Markkula M, Rossant J (1990) Embryonic stem cells alone are able to support fetal development in the mouse. Development 110:815–821

Nagy A, Rossant J, Nagy R, Abramow-Newerly W, Roder JC (1993) Derivation of completely cell-culture derived mice from early-passage embryonic stem cells. Proc Natl Acad Sci USA 90:8424–8428

Rossant J, Cross JC (2001) Placental development: lessons from mouse mutants. Nat Rev Genet 2:538–548

Schwenk F, Zevnik B, Bruning J, Rohl M, Willumeit A, Rode A, Hennek T, Kauselmann G, Jaenisch R, Kuhn R (2003) Hybrid embryonic stem cell-derived tetraploid mice show apparently normal morphological, physiological and neurological characteristics. Mol Cell Biol 11:3982–3989

Seibler J, Zevnik B, Kuter-luks B, Andreas S, Kern H, Hennek T, Rode A, Heinmann C, Faust N, Kauselmann G, Schoor M, Jaenisch R, Rajewsky K, Kuhn R, Schwenk F (2003) Rapid generation of inducible mouse mutants. Nucleic Acids Res 4:e12

Ventura A, Meissner A, Dillon CP, McManus M, Sharp PA, Van Parijs L, Jaenisch R, Jacks T (2004) Cre-lox regulated conditional RNA interference from transgenes. Proc Natl Acad Sci USA 28:10380–10385

Wang PJ, McCarrey JM, Yang F, Page DC (2001) An abundance of x-linked genes expressed in spermatogonia. Nat Genet 27:422–426

Wang ZQ, Kiefer F, Urbanek P, Wagner EF (1997) Generation of completely embryonic stem cell-derived mutant mice using tetraploid blastocyst injection. Mech Dev 62:137–145

4 Cloning The Laboratory Mouse by Nuclear Transfer

Kevin Eggan, Rudolf Jaenisch

4.1 Introduction

The development of the fertilized zygote into a complex organism has traditionally been thought of as a unidirectional process, with cells in the developing fetus becoming gradually more committed to a specific tissue type. The recent development of mammalian cloning by nuclear transplantation (NT) demonstrates that the mammalian oocyte has the remarkable ability to relieve the constraints imposed by cellular differentiation and return an adult nucleus to a totipotent, embryonic state (Wilmut et al. 1997; Gurdon and Colman 1999). Thus, cloning by NT provides a unique opportunity to elucidate the molecular and cellular mechanisms by which an adult cell can be returned to an undifferentiated state, a process termed nuclear reprogramming (Rideout et al. 2001).

Due to the prevalence of the laboratory mouse as a genetic and embryological model organism, it is a desirable system for the study of cloning and nuclear reprogramming. However, the mouse remains one of the more difficult mammals to clone, with transfer of NT technology from laboratory to laboratory occurring at a slow rate. With the technical difficulties of mouse NT in mind, we have assembled a chapter that describes the basic methodologies we use for the generation of cloned mouse embryos. The methodology we describe here is a modified version of the "Honolulu technique", first developed in the laboratory of R. Yanagimachi at the University of Hawaii; this technique relies on direct microinjection of the donor nucleus into the recipient oocyte (Fig. 4.1A,B) (Wakayama et al. 1998).

This chapter has three principal sections: (1) a discussion of the parameters influencing cloning efficiency and nuclear reprogramming, (2) a discussion of equipment set-up and methods for NT, and (3) an in-depth, step-by-step protocol for the NT procedure.

Kevin Eggan: The Stowers Medical Institute, Department of Molecular and Cellular Biology, Harvard University, Cambridge, MA 02138, USA, e-mail: Eggan@mcb.harvard.edu

Rudolf Jaenisch: Whitehead Institute for Biomedical Research and Department of Biology, Massachusetts Institute of Technology,
Cambridge, MA 02142, USA

Principles and Practice
Mammalian and Avian Transgenesis – New Approaches
Shirley Pease and Carlos Lois (Eds.)

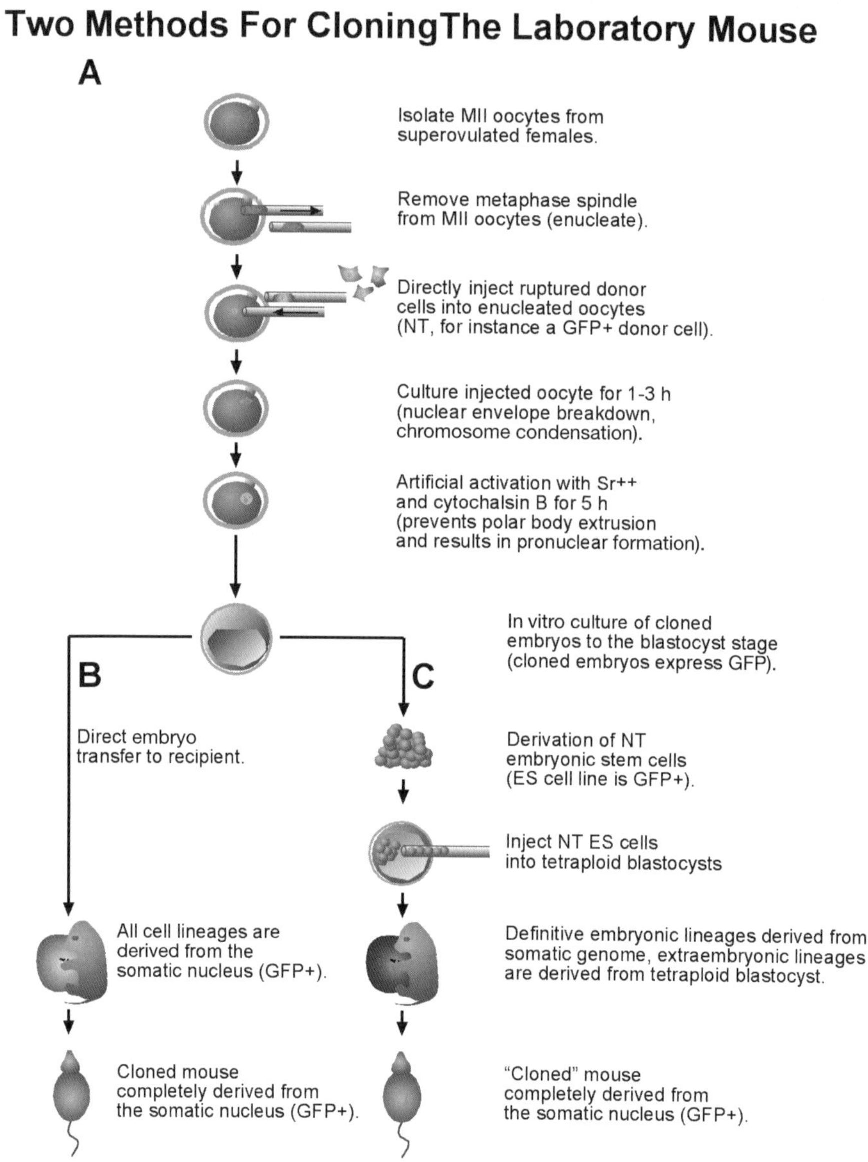

Fig. 4.1. A–C Cloning by direct injection nuclear transfer

In contrast to cloning procedures in other mammals, where the donor nucleus is introduced into the egg by cell-cell fusion, nuclear transfer in the mouse has been successful by the physical injection of an isolated nucleus into an enucleated egg. Therefore, we will concentrate on this procedure as used in mouse cloning. The chapter is written for an audience already skilled in handling and micromanipulating preimplantation mouse embryos, as required for the production of transgenic and mutant mice. For general information on embryo culture, mouse husbandry, tetraploid embryo complementation and embryo transfer surgery, we recommend Chaps. 4 and 11 of this book, and the more extensive descriptions that can be found elsewhere (Nagy et al. 2002).

4.2 Factors Influencing Cloning Success

A wide array of animals have been have been cloned through the introduction of somatic nuclei into enucleated oocytes (Campbell et al. 1996; Cibelli et al. 1998; Onishi et al. 2000; Polejaeva et al. 2000; Baguisi et al. 1999; Chesne et al. 2002; Wakayama et al. 1998; Shin et al. 2002). However, cloning by NT remains very inefficient, and cloned animals display a variety of embryonic, perinatal and postnatal phenotypes. These phenotypes include, but are not limited to, early developmental failures, dramatic fetal and placental overgrowth, neonatal respiratory failure, obesity, and premature death (Wakayama et al. 1998; Eggan et al. 2001; Tamashiro et al. 2002; Ogonuki et al. 2002). The biological basis of these phenotypes and the inefficiency of cloning by NT remain poorly understood.

Unfortunately, the universally low efficiency of NT experiments has made it difficult to dissociate technical difficulties from biological phenomena in cloning research. However, as research interests have shifted from whether cloning is merely possible to the parameters that influence its success, some of the variables that determine a successful cloning outcome have begun to emerge. To familiarize the reader with important technical aspects of cloning by NT, these parameters and the experiments that suggest they are critical will be described. Emphasis has been placed on results from the mouse, as it is the experimental system pertinent to this chapter. However, data from other animals are included to illustrate general issues of nuclear cloning.

4.2.1 Cell Cycle Status of the Donor Cell

Cloning experiments with a variety of cell types has demonstrated a correlation between the efficiency of NT embryo development to the blastocyst stage and the proportion of the donor cell population in the G1 phase of the cell cycle. When cumulus cells (Wakayama et al. 1998) and serum-starved fibroblasts (Wakayama and Yanagimachi 1999), primarily in a G1 state, were used for NT, the majority of activated embryos developed to the blastocyst stage. In contrast, when rapidly cycling embryonic stem (ES) cells were used as nuclear donors (Wakayama et al. 1999; Eggan et al. 2001), only a small percentage of NT embryos completed cleavage development. Consistent with the interpretation that differences in cell-cycle state caused this effect, culture conditions that force a higher proportion of ES cells into the G1 phase of the cell cycle, such as partial serum withdrawal, increase the potential of ES cell NT embryos to reach the blastocyst stage (Wakayama et al. 1999).

The importance of the donor nucleus cell-cycle state is linked directly to compatibility with the recipient oocyte cytoplasm. In the MII oocyte, metaphase/maturation promoting factor (MPF) levels are high (reviewed by Fulka et al. 1996). High MPF levels in the oocyte cytoplasm lead to somatic cell nuclear envelope breakdown and premature chromosome condensation after NT (Wakayama et al. 1998). The S phase of the cell cycle, characterized by DNA replication and a specific chromatin conformation, is likely to be incompatible with this condensation, leading to DNA damage and zygotic arrest. As most ES cells in a given population are in S phase, this is likely the cause of the poor in vitro development of ES-cell-derived embryos after NT.

4.2.2 Genetic Influences on the Cloning Process

When ES cells derived from several mouse strains were used as nuclear donors, inbred 129, C57/B6 and a variety of F1 ES cell lines all gave rise to newborn clones (Wakayama et al. 1999; Rideout et al. 2000; Eggan et al. 2001). However, clones from the inbred ES cell lines died shortly after birth due to respiratory failure (Wakayama et al. 1999; Rideout et al. 2000; Eggan et al. 2001). In contrast, most clones derived from five different F1 ES cell lines survived to adulthood (F1 here refers to ES cell lines derived from embryos produced by intercrossing parents with different inbred genetic backgrounds; (Rideout et al. 2000; Eggan et al. 2001)).

Neonatal lethality has also been reported in mice entirely derived from inbred ES cells injected into tetraploid blastocysts (Fig. 4.2) (Nagy et al.

A B C D

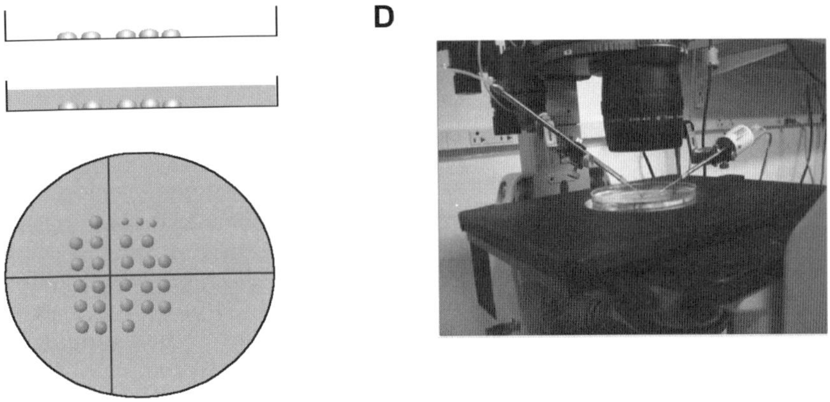

Fig. 4.2. A–D Setting up Petri-dish lid, micromanipulation chamber and micromanipulators, and inverted microscope set for piezo micromanipulation. Inverted microscope with Hofman modulation contrast set for nuclear transfer using the lid of a Petri dish as a micromanipulation chamber. Particularly, note the angle of inclination for the piezo micromanipulator, which is at about 20°

1990, 1993). Like inbred clones, ES cell-tetraploid neonates derived from inbred ES cell lines died shortly after delivery with signs of respiratory distress (Nagy et al. 1990; Eggan et al. 2001). In contrast, most ES cell-tetraploid neonates derived from six F1 ES cell lines, developed into fertile adults (Eggan et al. 2001). These results suggest that the death of inbred ES cell clones is not a direct result of the NT procedure itself but instead is due to the intrinsic character of inbred ES cells. It is possible that the decreased developmental potential and respiratory failure observed in inbred ES NT embryos could be due to delayed developmental timing relative to their F1 counterparts. However, experiments demonstrating that NT newborns derived from both F1 (Wakayama et al. 1998) and inbred 129 cumulus cells (Wakayama and Yanagimachi 2001) survive from birth to adulthood at equal frequency seem to argue against this conclusion.

It is possible that inbred ES cells may suffer some ill effects due to long-term cell culture, which ultimately leads to the death of ES-cell-derived offspring. Experiments suggesting that prolonged in vitro passage of ES cells can further aggravate these phenotypes in ES-cell-tetraploid-derived offspring lend further support to this hypothesis (Nagy et al. 1993). Interestingly, an F1 genetic background also seems to protect against these detrimental effects, since offspring from all F1 genotypes tested survived to adulthood, even at high passage (Eggan et al. 2001).

Cloned animals of other species, including sheep, also often display signs of respiratory distress at birth, especially following prolonged donor cell in vitro culture (McCreath et al. 2000). Thus, it could be that the ill-effects of long-term in vitro culture of donor cells, which are modified by genetic factors in mice cloned from ES cells, may be generally relevant to the survival of other cloned animals derived from cultured cells.

4.2.3 Intrinsic Developmental Potential of the Donor Cell

Comparing the efficiency of NT experiments using both somatic and embryonic cells has suggested that the intrinsic developmental potency of the donor cell plays a role in the cloning outcome. When mouse cumulus (Wakayama et al. 1998) and fibroblasts cells (Wakayama and Yanagimachi 1999) were used as nuclear donors, only 1–3% and 0.5%, respectively, of embryos transferred to surrogate mothers developed to term. In contrast, 5–25% of blastocysts generated by NT with ES cell (Wakayama et al. 1999; Rideout et al. 2000; Eggan et al. 2001) or blastomere (Tsunoda and Kato 1997) nuclei survived until birth.

These observations have sparked debate as to whether all nuclei have the capacity to direct embryogenesis after NT. One viewpoint is that surviving clones are derived from rare cells with intrinsic developmental potential that are present at a low frequency in the donor cell population, such as somatic stem cells (Clarke et al. 2000; Hochedlinger and Jaenisch 2002b). Thus, the great inefficiency of cloning might reflect the rare nature of donor cells with the developmental capacity to direct development after NT. Experiments using mature lymphocytes (Hochedlinger and Jaenisch 2002a) and neurons (Eggan et al. 2004) as nuclear donors have now demonstrated that bona fide differentiated cells can give rise to ES cell lines and cloned mice. However, in both of these cases a two-step cloning process including an ES cell intermediate was used (Fig. 4.1C), raising the question of whether complete reprogramming of a terminally differentiated nucleus requires passage of the genome through the ES cell state. It is possible that the embryo and the oocyte alone can directly reprogram the epigenetic state of a terminally differentiated cell to give rise to a cloned embryo after direct embryo transfer, but this may be exceedingly inefficient and has not been achieved as yet (Fig. 4.1B).

4.2.4 Cellular Identity of the Donor Cell

Several lines of evidence suggest that the identity of the donor cell nucleus may influence both gene expression and phenotypes in cloned embryos and animals. First, in classical NT experiments using the amphibian Rana pipiens it was observed that, when endoderm nuclei were used as nuclear donors, cloned embryos showed normal development of endodermal derivatives but abnormal development of mesodermal and ectodermal tissues (Briggs and King 1952, 1957). Complementary phenotypes were observed in cloned embryos derived from neural ectoderm nuclei, and these phenotypes were subsequently termed endoderm and ectoderm syndrome (DiBerardino and King 1967).

Similarly, there is mounting evidence suggesting the donor cell type might influence the phenotype of cloned mice. When cumulus cells were used as nuclear donors, many cloned mice became obese (Tamashiro et al. 2002). In contrast, mice derived from sertoli cells of the same genetic background did not become obese, but instead died prematurely with some signs of tumorigenesis and kidney failure (Ogonuki et al. 2002). Our own analysis of placental transcriptional profiles from cloned mice revealed abnormalities in gene expression that were common to all cloned animals. However, we also detected gene expression abnormalities that were present in all animals cloned from cumulus cells but that were not found in animals cloned from ES cells, which had their own set of common abnormalities that were not found in cumulus clones (Humpherys et al. 2002). Together, these results suggest there may be incomplete cell-type specific reprogramming of the donor nucleus, and that these cell-type specific effects may have important influences on the development and phenotypes of cloned mice.

4.2.5 Epigenetic Reprogramming after Nuclear Transfer

The most intriguing issue in cloning by NT is the problem of epigenetic reprogramming (Gurdon and Colman 1999). In order for clones to complete development, genes normally expressed during embryogenesis, but silent in the somatic donor cell, must be reactivated. To date, the efficiency of deriving live cloned animals has been low, independent of the cell type used as nuclear donor, with two notable exceptions. Nuclei isolated from ES cells and from embryonic blastomeres generated viable cloned animals with a significantly higher efficiency than any somatic donor cell type (reviewed in Rideout et al. 2001). As stated above, this observation is consistent either with the notion that only a limited number of somatic cells, those with sufficient developmental potential, are competent nuclear donors, or that the genome of pluripotent embryonic cells is more easily reprogrammed than that of somatic cells.

In normal development, reprogramming of the genome occurs during gametogenesis, a complex process that assures that, when combined at fertilization, the genome of the two gametes can faithfully activate early embryonic genes (Barton et al. 1984; Kafri et al. 1992). In cloning, reprogramming presumably must occur in the short interval between transfer of the donor nucleus into the egg and the onset of cellular differentiation at the blastocyst stage, a cellular context dramatically different from that of normal fertilization. The challenge has been to identify the epigenetic abnormalities, arising either as a result of faulty reprogramming or during donor cell in vitro cultivation and aging, that are responsible for the inherently inefficient nature of cloning and the phenotypes of cloned animals. The hope has been that by understanding these abnormalities we might determine the nature of epigenetic information that either is, or is not, reprogrammed after NT, and that we might use this information to pinpoint the time at which reprogramming occurs, allowing identification of the molecular machinery responsible.

A major focus of this research has been to determine the identity of epigenetic information that either is, or is not, reprogrammed after NT. For instance, X chromosome inactivation and telomere length adjustment are rapidly and robustly reprogrammed after NT, indicating that disturbances in these epigenetic states are not likely impediments to the development of clones (Eggan et al. 2000; Tian et al. 2000; Betts et al. 2001; Lanza et al. 2000; Young et al. 1998). Epigenetic information encoding mono-allelic expression of imprinted genes, however, is not restored to a functional state after NT, and disturbances in the expression of these genes may lead to the severe over-growth observed in many cloned animals (Bartolomei and Tilghman 1997; Humpherys et al. 2001; Inoue et al. 2002). Finally, emerging evidence suggests that the reactivation of developmentally regulated gene expression may occur incorrectly, or incompletely, after NT, hinting at a root cause for the early developmental arrest of many cloned embryos (Boiani et al. 2002; Bortvin et al. 2003).

4.3 Methods, Equipment and Techniques

4.3.1 Embryo Culture Media and Common Stock Solutions

Preimplantation embryos are very sensitive to organic solvents, detergents and perturbations in pH, thus we recommend using disposable plastic or dedicated glassware for media preparation. We use ultrapure H_2O, such as that available from Specialty Media (Phillipsburg, NJ). KSOM or CZB embryo culture media used for NT are produced using a master salt mix as a base (Chatot et al. 1990).

4.3.2 Mouse Strains and Animal Husbandry

We purchase 4- to 6-week-old B6D2F1 female oocyte donors from Charles River labs (www.criver.com). These females are housed in our facility for a minimum of 1 week before superovulation. To produce pseudopregnant recipients for embryo transfer, we mate Swiss Webster females, weighing between 28 and 40 g, to vasectomized Swiss Webster males. Mice are housed on a 12 h day/night cycle. (For additional detail, see Chap. 11 on Ancillary Techniques, this manual.)

4.3.3 Preparation of Cumulus Cells for Nuclear Transfer

Cumulus cells for NT are isolated from cumulus complexes in parallel with MII oocytes as previously described ((Wakayama et al. 1998); also see below). After 5 min of hyaluronidase treatment, cumulus cells are aspirated into a transfer pipette with a minimal amount of medium and deposited in a 500 μl drop of HCZB under mineral oil in a small dish. This dish is then placed on ice until needed for NT.

4.3.4 Preparation of Tail-tip Cells for Nuclear Transfer

We prepare tail-tip donor cells as described for the production of the first male cloned animals (Wakayama and Yanagimachi 1999). Euthanise the donor mouse by approved means and then amputate one-half of the tail. After amputation, the biopsy is placed into Wescodyne for 2 mins, washed several times through Hepes-buffered saline (HBS) and the outer layer of skin is removed. After several more washes through HBS, and finally through Dulbecco's modified Eagle's medium (DMEM) containing 15% fetal calf serum (FCS), the tail is chopped into very small pieces with a scalpel on the lid of a Petri dish. The pieces of tail are then split equally among two to three wells of a six-well tissue culture dish. The wells of the tissue culture dish are then filled with DMEM supplemented with 15% FCS. After 3 days of culture in 5% CO_2 at 37 °C the medium is changed and fibroblasts should be observed growing on the bottom of the dish. These cultured cells are best used 1–2 weeks after generation. Around 24–48 hour prior to NT, DMEM with 15% FCS is replaced with DMEM with 0.05% FCS in order to force the donor cells to exit the cell cycle. To prepare tail-tip cells for NT, they are trypsinized, washed twice in DMEM with 15% FCS, and placed on ice until needed.

4.3.5
Culture and Preparation of ES donor cells for Nuclear Transfer

ES cell culture is carried out essentially as previously described (Hogan et al. 1994). ES cells are cultured in ES cell medium [ESCM; DMEM with 15% fetal calf serum (Hyclone), 0.1 mM non-essential amino acids (Gibco, Rockville, MD), 2 mM L-glutamine, 50 IU Penicillin, 50 IU Streptomycin (Gibco) and 0.1 mM beta-mercaptoethanol (Sigma, St. Louis, MO), 1,000 U/ml leukemia inhibitory factor (LIF)] on gelatinized tissue culture ware (Falcon; www.bdbiosciences.com) pre-plated with a monolayer of gamma-irradiated primary mouse embryo fibroblasts (MEFs). Recently we have begun supplementing our ES cell media with the MEK kinase inhibitor PD98059 (Cell Signaling Technology, www.cellsignal.com) at a final concentration of 50 μM; this compound has been reported to inhibit differentiation of mouse ES cells (Burdon et al. 1999). For nuclear transplantation, ES cells are cultured in ESCM with 5% FCS for 24–48 h to slow down their cell cycle. Following starvation, ES cells are trypsinized, resuspended in DMEM and pre-plated on a standard 10 cm tissue culture dish for 30 min to remove feeder cells and debris. Following pre-plating, trypsinized ES cells are washed once in ESCM and placed on ice until NT.

4.3.6
Microscope Set-up

All micromanipulation required for NT is carried out with standard hydraulic micromanipulators (Narishige, M188NE or equivalent, www.narishige.co.jp) on an inverted microscope (such as Nikon TE200) with Hofman modulation contrast optics (Fig. 4.2). The lid of a Petri dish is routinely used as a micromanipulation chamber (Fig. 4.2). For NT by direct injection – the "Honolulu" method – we use the piezo micromanipulator made by Primetech [Ikabari, Japan, distributed in the United States by Brinkmann (www.brinkmann.com)]. To cope with the significant back-pressure generated during piezo microinjection, we recommend the IM6–2 microinjector (Narishige). This manipulator is also suitable for the precise micromanipulation procedures required for successful NT.

4.3.7
Micromanipulation Instruments for Nuclear Transfer

Flat-tipped microinjection pipettes are used for NT. The method used to produce microinjection instruments is identical to that used to produce instruments for piezo-assisted blastocyst injection with ES cells, as described in Chap. 3. For enucleation, a pipette with an internal diameter of 8 μm is used. For NT, a microinjection pipette just smaller than the diameter of the donor cell is used (ES cells and cumulus cell 6 μm, tail-tip cell 8–10 μm). For NT, 3–5 μl mercury is loaded into the back of the pipette (a 1 ml syringe fitted with silicon tubing is appropriate for this task). Great care should be taken while back loading mercury into the pipette as the back pressure generated is substantial and can cause the mercury to be expelled suddenly from the back of the pipette. After the pipette has been loaded with mercury it is affixed to the pipette holder (Fig. 4.2).

4.3.8 Isolation of Metaphase II Oocytes for Nuclear Transfer

A detailed description of superovulation and microdrop embryo culture can be found elsewhere (Hogan et al. 1994). Briefly, all embryo culture should be carried out in 20–30 μl drops of medium (unless otherwise noted) on a standard Petri dish flooded with mineral oil (see Fig. 4.2). Ovulated MII oocytes for NT are isolated from super-ovulated 8- to 10-week-old B6D2F1 females. Superovulation is induced by intraperitoneal (IP) injection of 5 IU pregnant mare's serum (PMS) (Calbiochem, San Diego, CA) performed between 6 and 7 pm; 46–48 h after PMS, 5 IU human chorionic gonadotropin (HCG) (Calbiochem) are injected IP.

At 14–15 h following the administration of HCG, female oocyte donors are sacrificed and the oviducts removed. After dissection, oviducts can be placed directly into 200 μl drops of HCZB placed on a Petri dish and covered with mineral oil. One of the droplets is supplemented with bovine testicular hyaluronidase at a final concentration of 0.1% w/v (Sigma). The oocytes are isolated from the oviducts by tearing the ampullae with forceps, releasing small drops of oviductal fluid containing the oocytes into the mineral oil. Once all cumulus complexes have been removed from the oviducts they can be quickly and simultaneously moved into the drop of HCZB containing hyaluronidase. After 2–3 min, the hyaluronidase will begin to dissociate the cumulus complexes; 5–10 min after the oocytes were placed in hyaluronidase, they should be washed, using a mouth-controlled aspirator assembly, through the other drops of HCZB medium, eliminating any hyaluronidase and remaining cumulus cells.

Following cumulus cell removal, oocytes should be transferred into MCZB medium that has been pre-equilibrated for 20–30 min at 37 °C under 5% CO_2 in air. Alternatively, we now also use KSOM with amino acids (Specialty Media) for long-term embryo culture. Oocytes should be washed through several drops of medium to remove residual HZCB. Recipient oocytes should be used for enucleation within 2 h of isolation.

4.3.9
Enucleation of MII Oocytes

After their isolation, groups of 15–20 oocytes are placed on the microscope stage in a micromanipulation chamber prepared as described in Fig. 4.2, containing HCZB with 5 μg/ml cytochalasin B (Sigma). The oocytes should rest in cytochalasin-containing medium for 5–10 min to allow actin-cytoskeletal depolymerization to occur. Cytoskeletal depolymerization allows the metaphase spindle to be removed from the oocytes with minimal chance of lysis. Following cytoskeletal depolymerization, enucleation can be performed as described in Fig. 4.3. After enucleation, oocytes should be washed through several drops of KSOM or MCZB and returned to one of these media in the incubator. This process can then be repeated until the desired number of enucleated oocytes has been collected. Enucleation of a single group of embryos should take no longer than 15–20 min.

4.3.10
Nuclear Transfer

For nuclear transfer, donor cells are diluted 1:10 into a drop of HCZB containing 11% w/v polyvinylpyrrolidone (PVP). It is critical that the cells are mixed thoroughly with the PVP to ensure that they are not damaged by the ionic gradient between the two media with and without the volume-excluding PVP. Just before donor nucleus isolation, a group of 10–20 enucleated oocytes should be placed on the stage in HCZB containing 5 μg/ml cytochalasin B. Donor nuclei are isolated by aspirating intact cells into the injection pipette and slowly working them back and forth in the needle, while applying gentle pulses with the piezo micromanipulator. In general, 3–10 nuclei are isolated for NT at one time. Following donor nucleus isolation, the instruments are moved to the drop of medium containing the enucleated oocytes and direct injection NT can be performed as described in Fig. 4.3.9. After NT, reconstructed embryos are washed through several drops of KSOM or MCZB and returned to the incubator.

4.3.11
Oocyte Activation and Subsequent Culture of Cloned Embryos

We activate reconstructed NT embryos 1–3 h after NT in Ca^{++}-free MCZB medium containing 10 mM Sr^{++} and 5 μg/ml cytochalasin B. Oocytes are

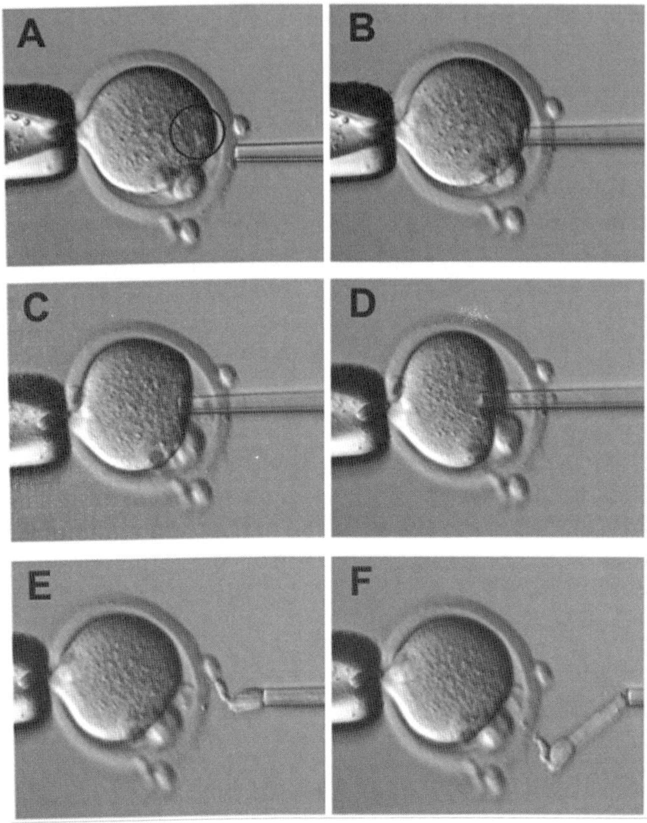

Fig. 4.3. A–F Piezo micromanipulator enucleation of metaphase II (MII) mouse oocyte. **A** Mouse MII oocyte immobilized with a standard holding pipette. The MII spindle has been circled. Note that the spindle has an optical nature different from the rest of the oocyte and can be seen as a more refractive portion of the cytoplasm. **B** Application of piezo micromanipulator combined with gentle suction to the enucleation pipette leads to zona "drilling", which allows access of the blunt needle to the perivitelline space. **C** Before aspirating the spindle into the enucleation pipette it is often worthwhile to touch the spindle with the end of the pipette. If the spindle moves, one can be confident that the needle is in the correct location and that the spindle will be removed with a minimum of cytoplasm. **D** After the pipette is in the correct location, suction is applied to the enucleation needle, leading to aspiration of the spindle into the pipette. Note that the refractive portion of the cytoplasm is being drawn into the pipette. **E** Once the spindle is two-thirds to three-quarters inside the pipette, the pipette should be drawn away from the oocyte. If done properly the spindle will remain in the pipette, removing the minimum of cytoplasm. **F** To confirm that enucleation was complete, the spindle can be observed after expulsion from the enucleation pipette. Unlike the rest of the ooplasm, which will form a sphere when discarded from the needle, the spindle has a rigid character and will remain in a bar-shaped karyoplast. Often the karyoplast will have a distinct "dog-bone" appearance

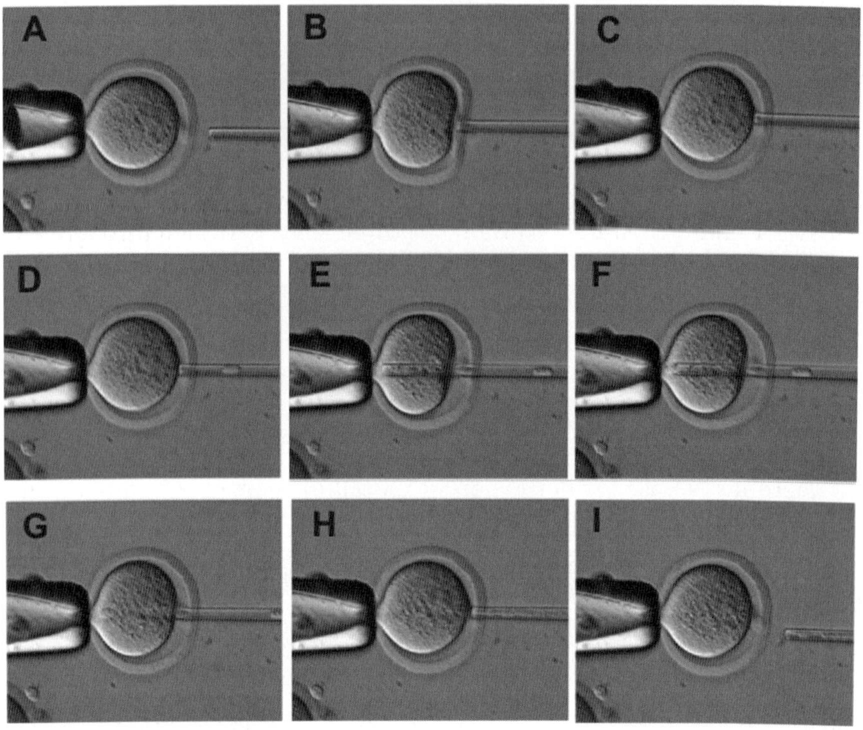

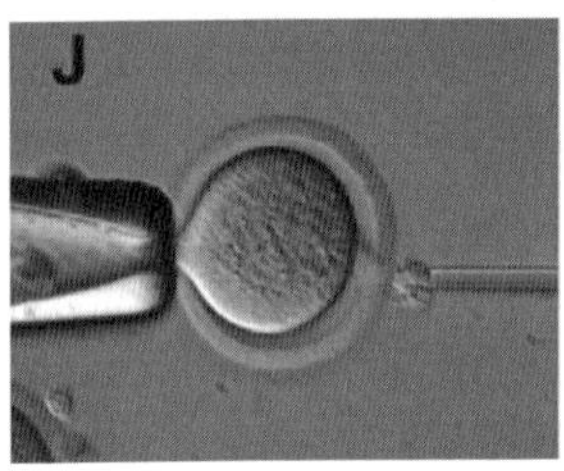

Fig. 4.4. A–J Piezo micromanipulator direct injection nuclear transfer with hole removal method. **A** After cell pickup, the instruments are returned to the culture drop containing enucleated oocytes. The oocyte to be injected should be immobilized with the largest portion of the perivitelline space facing the injection pipette. **B** The blunt injection pipette cannot enter the perivitelline space without use of the piezo. **C** Once the piezo element is applied, the zona can be drilled. Zona drilling will cause a plug of zona to be aspirated into the needle. After drilling, this piece of zona should be expelled into the perivitelline space. **D** Expelling medium into the perivitelline space brings the donor nucleus towards the oocyte. **E** Just before the donor nucleus reaches the end of the pipette, the pipette should be moved forward into the oocyte, forming a channel of ooplasmic membrane around the injection pipette. **F** A brief pulse of the piezo breaks the oolema, creating a small hole, allowing the donor nucleus to be deposited inside the oocyte. Note nucleus just outside of injection pipette, proximal to the opening of the holding pipette. **G** Immediately after nucleus drop off, the needle should be quickly withdrawn just to the opening of the channel created in the membrane. **H** Once at the opening of the channel, suction should be applied. Proper suction will aspirate the membrane into the injection pipette. If this is done correctly, the hole created by microinjection is aspirated into the needle. **I** If the needle is removed carefully and quickly, the fluid ooplasmic membrane with pinch off from the end of the needle, topologically removing the hole from the membrane! **J** After hole removal, the remaining cytoplasm should be expelled from the injection needle

activated under these conditions for 5.5–6 h. It is critical that NT embryos be carefully and thoroughly washed in this medium prior to incubation, to remove the calcium present in KSOM. Failure to wash embryos can result in lysis. We generally culture a maximum of five NT embryos per 20 μl drop of activation medium. Oocyte death during activation, which can lead to Sr^{++} precipitation and further oocyte death, is not uncommon. Keeping a minimal number of oocytes in each drop lowers this risk to any particular NT embryo. After activation, NT embryos are washed through several drops of KSOM or MCZB. Following washing, embryos are cultured in KSOM or MCZB until embryo transfer.

4.3.12 Derivation of Nuclear Transfer ES Cells

The methods we use for derivation of ES cell lines from NT embryos are similar to those previously reported (Hogan et al. 1994) with several modifications (Hochedlinger and Jaenisch 2002a; Eggan et al. 2004). Following oocyte activation, NT embryos are cultured in vitro in KSOM for 3–4 days. Embryos reaching the morula or blastocyst stage after this time are used for ES cell derivation. NT morulae and blastocysts are exposed to acid Tyrode's solution (Specialty Media) to remove the zona pellucida and then transferred into individual wells of a four-well dish pretreated with 0.1% gelatin and seeded with gamma-irradiated primary MEFs. The embryos are cultured in standard ESCM (see above) supplemented with PD98059 (Cell Signaling Tech, www.cellsignal.com). PD98059 is a MEK kinase inhibitor, which, in addition to inhibiting the differentiation of mouse ES cells, increases the efficiency with which ES cell lines are derived.

Once the zona-stripped embryos are placed on MEFs they are left undisturbed for 48 h. After 48 h, the embryo explants must be monitored every day. At 72 and 96 h following explantation, a few drops of ESCM with PD98059 are added to each well of the dish. Once the embryos attach to the dish, one-half of the medium can be removed from the well and replaced with fresh medium every 24 h. Special attention should be paid to not switching the medium too soon as this can result in embryo loss. Attached embryos should be monitored every 24 h to assess outgrowth of the inner cell mass (ICM). In our experience, ICM outgrowth can take much longer for NT embryos than for fertilized embryos, and can be observed anytime between 5 and 15 days following embryo explantation. Actively growing ICMs should be picked with a mouth pipette, rinsed in hepes buffer and dispersed in 0.1% trypsin-EDTA for 10 min. Following physical dispersal of the picked ICM, cells should be plated on fresh MEFs in ESCM with PD98059. Initial ES cell colonies can often be observed 48–

72 h following ICM dispersal. Following derivation, NT ES cell lines are maintained by standard methods.

4.3.13 Embryo Transfer of Cloned Embryos

For production of cloned mice, we have had best success transferring NT embryos at the two-, four- or eight-cell stage to the oviducts of 0.5 days post coitum (dpc) pseudopregnant females. Although it does work in our hands, we have not found transfer of NT morula/blastocyst stage embryos to the uteri of 2.5 dpc pseudopregnant as consistently successful as oviduct transfer (Wakayama et al. 1998). The oviduct transfer method we use is a variation of that described elsewhere (Hogan et al. 1994, Chap. 11 of this manual). For oviduct transfer, mice are anesthetized with the barbiturate Avertin. In the meantime, NT embryos are retrieved from the incubator and moved from microdrop culture under mineral oil, to microdrop culture under air. This transfer prevents oil droplets, which may cling to the transfer pipette, from entering the reproductive tract. To access the oviduct, a small incision is made one inch ($\sim$ 2.5 cm) from the dorsal midline, at the clear depression in the back of the animal. After cutting through the skin, another incision is made in the body cavity, being careful to avoid red blood vessels and white nerves. The ovary and oviduct should be clearly visible (see Hogan et al. 1994, for a diagram of the ovary and oviduct, and Chap. 11 of this manual). After the second incision is made, a pair of forceps can be used to reach into the body cavity and carefully withdraw the ovary and oviduct. Under increased magnification it should be possible to locate the end of the oviduct as it enters the ovarian bursa. Once the end of the oviduct has been identified, immobilize the ovary with a haemostat.

To transfer the embryos, pick up 5–10 μl embryo culture medium in the transfer pipette, followed by two small air bubbles and then 10–15 NT embryos in a similar volume of medium. Holding the transfer pipette and a syringe with a 30.5 gauge needle in the same hand, use the needle to make a small hole in the oviduct wall, just outside the ovarian bursa. Special effort should be made not to rupture blood vessels in the oviduct. After making the hole, drop the syringe, adjust your grip on the transfer pipette and place the tip of the transfer pipette into the hole made by the syringe. Lastly, blow the embryos and the air bubbles into the oviduct. The air bubbles and medium should push the embryos into the swollen ampulla. After transfer, carefully replace the ovary back into the body cavity, close the hole in the body wall with a single suture, and staple the wound closed with two or three autoclips. Repeat on the other oviduct or proceed to another recipient. Note that we find this transfer method less invasive and more reliable then the oviduct transfer method described elsewhere (Hogan et al. 1994) where the bursa is ripped from the ovary and the embryos introduced via the infundibulum.

4.3.14 Cesarean Section and Cross Fostering of Cloned Animals

We routinely carry out cesarean section in the morning of recipient 19.5 dpc (see Chap. 3, section 3.3.10). After sacrifice of the recipient, full-term pups are quickly but carefully removed from the uterus, the umbilicus is carefully cut from the pups and the placentas arrayed in such a way that their identities can be maintained. After removal from the uterus, neonates are quickly wiped off with a Q-tip or Kim-wipe, paying special attention to airways. Pups are kept under a heat lamp until cross-fostering. Generally we prefer to cross-foster to BALB/c foster mothers that have delivered the previous day. To cross foster, remove the BALB/c litter from the cage with a little bedding. Mix pups to be fostered and bedding thoroughly until pups have taken on the smell of the bedding. We generally leave 6–8 pups per foster mother, i.e., number of NT pups to be fostered + X BALB/c pups = 6–8 in total.

4.4 Protocol for Direct Injection Nuclear Transfer

4.4.1 Production of Embryo Culture Medium, Reagents and Mice

4.4.1.1 Master Salts

- Begin with 980 ml ultrapure H_2O in a sterile 1 l bottle.
- Add dry components:

NaCl	4,760 mg	Sigma S-5886
KCl	360 mg	Sigma P-5405
$MgSO4 \cdot 7H_2O$	290 mg	Sigma M-2773
KH_2PO_4	160 mg	Sigma P-5655
EDTA 2NA	40 mg	Sigma E-6635
Glucose (D)	1,000 mg	Sigma G-6152

- Add liquid components:

Na-lactate (lactic acid)	5.3 ml	Sigma 44263
Pen'Strep (100x TC)	10 ml	Gibco 15140–122

- Two primary media types are made from these master salts: bicarbonate-buffered MCZB medium for long-term culture at 37 °C with 5% CO_2, and HCZB medium for short-term culture and manipulation at room temperature in air. To prepare these media, the master stock is subdivided and supplemented to produce MCZB and HCZB stocks salts.

4.4.1.2
MCZB Stock Salts

- Sterile filter 500 ml master salts into a sterile container.
- MCZB stocks can be used for up to 3–4 months if stored at 4 °C.

4.4.1.3
HCZB Stock Salts

- Start with 500 ml master salts.
- Add 50 mg PVA (cold-soluble; Sigma P-8136).
- Stir for 30–60 min and sterile filter.
- Store for up to 3 months at 4 °C.
- These CZB stocks are used as a base for MCZB and HCZB media, which must be prepared every 2 weeks and should be stored at 4 °C.

4.4.1.4
MCZB For Long-Term Embryo Culture in 5% CO_2

- Start with 99 ml MCZB stock salts.
- Add:

$NaHCO_3$	211 mg	Sigma S-5761
Na-pyruvate (pyruvic acid)	3 mg	Sigma P-4562
L-Glutamine	15 mg	Sigma G-8540
Bovine serum albumin (BSA)	500 mg	Sigma A-3311

- Add liquid:

128 mM $CaCl_2$	1 ml	Sigma C-7902

- Swirl until dissolved, sterile filter. KSOM medium purchased from Specialty Media can be substituted for MCZB.

4.4.1.5
Ca^{++}-Free MCZB For Oocyte Activation in 5% CO_2

- Start with 99 ml MCZB stock salts.
- Add:

$NaHCO_2$	211 mg	Sigma S-5761
Na-pyruvate (pyruvic acid)	3 mg	Sigma P-4562
L-Glutamine	15 mg	Sigma G 8540
BSA	500 mg	Sigma A-3311

- Swirl until dissolved, sterile filter.

4.4.1.6
HCZB (use for micromanipulation in air)

- Start with 99 ml HCZB stock salts.
- Add:

Hepes-Na	520 mg	Sigma H-3784
$NaHCO_3$	42 mg	Sigma S-5761

- Add liquid:

128 mM $CaCl_2$	1 ml	Sigma C-7902

- adjust pH to 7.5 with 1 N HCl.
- Swirl until dissolved, sterile filter.

4.4.1.7
Cytochalasin B (100× stock)

500 μg/ml cytochalasin B (Sigma) in DMSO. Store at −20 °C.

4.4.1.8
Strontium chloride (10× stock)

100 mM mM SrCl in H_2O. Store at room temperature.

4.4.1.9
HCZB with 11% w/v PVP

For nuclear transfer, donor cells are suspended in an HCZB solution containing 11% w/v PVP. We have found that the composition and preparation of this solution is one of the most critical factors in cloning success. HCZB medium (20 ml) should be placed in a 50 ml conical tube or beaker. Place 2.2 g PVP (MW 360,000, ICN Biochemicals, Aurora, OH) on top of the liquid. The conical tube should then be closed or the beaker sealed, placed at 4 °C and left undisturbed for 72–96 h. After 96 h, the PVP will have entered solution. The HCZB/PVP solution should then be passed through an 8 μm filter and stored at 4 °C.

4.4.1.10
Mice

Mice should be superovulated with PMSG (3 days earlier) and HCG (1 day before NT) but not mated. For instance, PMSG Friday afternoon (∼ 6:00 p.m.), HCG Sunday afternoon (∼ 6:00 p.m.), egg collection Monday morning 8:00–9:00 a.m. (+14–15 h after HCG, +16 h is too late).

4.4.2
Oocyte (Egg) Collection

Prepare Petri dishes 1–5 in advance.

- Petri dish 1:
  - One drop of HCZB with hyaluronidase (HCZB is based on HEPES buffer and does not require CO_2 for pH-equilibration).
  - Several (four) drops of HCZB (to wash eggs after hyaluronidase).
  - Cover with oil.
- Petri dish 2:
  - Many small drops of KSOM (20 μl) to wash hyaluronidase/HCZB from oocytes. One or two large (100 μl) drops of KSOM in which to keep washed oocytes.
  - Many small drops of KSOM to keep enucleated oocytes.
  - Cover with oil, place at 37 °C.
- Petri dish 3 (enucleation and nuclear transfer dish):
  - For this chamber use the lid of a Petri dish.
  - Draw four quadrants on the outside of the lid (Fig. 4.2).
  - Mix 10 μl cytocholasin B (100× stock in DMSO) with 990 μl HCZB in a microcentrifuge tube; this is enucleation and microinjection medium.
  - Make many small drops of enucleation medium in three out of the four quadrants.

- In the remaining quadrant, make several small drops of HCB-PVP medium (prevents needles from becoming clogged, also to use for isolation of nuclei).
- Cover with oil, place on the micromanipulation stage.

• Petri dish 4 (culture dish for oocyte activation):
  - Prepare activation medium:

10 μl	Cytocholasin B (1:100 dilution, use once)
100 μl	100 mM Sr^{2+} (10× stock)
890 μl	Ca^{2+}-free MCZB
1,000 μl	Total volume

  - Place two large drops activation of medium on the dish. These will be used to wash KSOM from reconstituted eggs before activation.
  - Use many small drops of activation medium for NT embryo culture at activation stage. Culture 5–10 reconstituted embryos per drop.
  - Several drops of KSOM-to wash eggs after activation (to free from cytocholasin B, Sr^{2+} etc.).

• Petri dish 5 (long term culture):
  - Make several drops of KSOM or MCZB medium for long-term culture.
  - Cover with oil, place at 37 °C.

4.4.2.1
Isolate ovaries and oocytes

1. Sacrifice female oocyte donors by accepted method and dissect out oviducts.
2. Put oviducts in oil (dish 1), tear ampula(e), eggs spill into oil (oocytes surrounded by cumulus cells).
3. Simultaneously, push cumulus/oocyte complexes into hyaluronidase.
4. Let oocytes rest at room temperature 2–5 min (maximum 10 min) until cumulus cells start to fall away.
5. Wash oocytes through HCZB (3–4 times) to remove hyaluronidase.
6. Move oocytes to dish 2 and wash through two KSOM drops to remove HCZB.
7. Place dish 2 back in 37 °C incubator.

4.4.3
Enucleation

Note: Before starting, make sure the micromanipulator lines are filled with H_2O. Air bubbles in the micromanipulation lines will prevent the manipulators from responding properly.

4.4.3.1
Setup

1. Fill piezo pipette needle with mercury:
   - Pipette 3–5 μl mercury into an insulin syringe fitted with a piece of silicon tubing in place of the needle.
   - Use the syringe to backfill the enucleation pipette with mercury. The mercury should be advanced nearly to the tip of the enucleation needle.
2. Attach the needle to the right-hand micromanipulator:
   - Position the instrument so that it can be lowered into the drop of HCZB-PVP medium.
   - Carefully lower the enucleation pipette into the PVP medium.
   - See Fig. 4.2.
   - Using the microinjector syringe, expel microscopic drops of mercury from the small tip of the enucleation needle and then aspirate in PVP.
   - Pipette the mercury up and down in the enucleation pipette (this will clean the pipette and coat the needle with PVP, preventing stickiness on its inner and outer surfaces.
   - After washing the pipette with PVP, rinse the outside of the needle through a drop of injection/enucleation medium while retaining PVP medium inside the pipette.
   - PVP is toxic to the oocytes in large quantities and smaller amounts of PVP may affect egg viability. Therefore, avoid moving PVP on the instruments directly into drops of enucleation/injection medium containing oocytes.
   - Once the enucleation pipette is in place, the holding pipette can be similarly brought into position. Never expose the holding pipette to the medium containing PVP. Its large surface area is difficult to rinse

and can introduce large amounts of the chemical into microdrops that contain oocytes.

- Turn on the piezo drill and set the instrument at a moderate power for "zona drilling" (speed 3–5, intensity 3–5).

4.4.3.2 Begin enucleation

1. For beginners, retrieve 5–15 eggs from dish 2 and place them on the microscope stage in a drop of enucleation medium on dish 3.
2. Let eggs rest for 5–7 min in enucleation medium before beginning micromanipulation.
3. Cytocholasin B disrupts the actin cytoskeleton within the egg, allowing the ooplasm and membrane to move more fluidly. This will allow enucleation without lysis of the oocyte.
4. Do not keep oocytes on the stage for more than 20–25 minn. Oocytes should be manipulated and returned to the incubator within this time-frame.
5. Under a 20× or 40× objective, identify the metaphase spindle of the oocyte.
6. See Fig. 4.3 for photographic depiction of enucleation.
7. Rotate the egg with the injection pipette until you see the differently refractive metaphase spindle. Depending on the orientation of the egg, the spindle will appear either as a circular structure or as a blunt rod.
8. Position the oocyte on the holding pipette so that the spindle and the zona are in the same focal plane, with the spindle positioned close to the enucleation pipette.
9. "Drill" through the zona pellucida using the piezo micromanipulator.
10. Ensure there is sufficient gap between the zona and egg membrane, otherwise the pipette may lyse the egg.
11. Aspirate the metaphase plate into the pipette without taking much of the cytoplasm. A good rule of thumb here is to not begin aspirating until you can see that the pipette can move the spindle.
12. Once 60–70% of the spindle is in the pipette, slowly pull the needle away from the oocyte. Rapid movement or aspiration of the spindle into the needle too quickly can cause oocyte lysis.

13. As the pipette is pulled away from the oocyte, the fluid membrane will close behind it. The membrane is never compromised.
14. The spindle is more rigid than the rest of the ooplasm. One guideline for whether the spindle has been successfully removed is to expel the karyoplast from the enucleation pipette. When the spindle karyoplast is expelled from the needle it will be more rigid then cytoplasm alone and will often take on a characteristic "bone" shape.
15. Use the smallest enucleation needle possible, thus reducing the amount of cytoplasm taken up with the karyoplast during enucleation.
16. When finished with all oocytes, or after 20–25 min, move enucleated oocytes back into KSOM in Petri dish 2.
17. Wash enucleated eggs through three consecutive drops of KSOM.
18. Place enucleated oocytes in a fourth drop of KSOM.
19. Repeat enucleation in batches of oocytes that can be completed within these time constraints, keeping un-enucleated and enucleated oocytes in dish 2 at 37 °C.
20. Complete enucleation after no longer than 2 h.

4.4.4 Donor Nucleus Isolation

1. Remove enucleation needle.
2. Fill injection/NT needle with mercury as for enucleation pipette.
3. Attach to manipulator and lower into PVP as described for the enucleation needle.
4. Using a mouth-controlled pipette or a P20 Pipetteman (Gilson), place a few thousand donor cells into a clean drop of PVP medium.
5. Leave the remaining donor cells on ice.
6. Mix cells thoroughly with PVP. Cells may begin to die after an hour and a new drop of donor cells should frequently be made, particularly if the cells become sticky.
7. Place a group of oocytes into a drop of enucleation medium (again a number of oocytes that can be manipulated in less the 25 min).
8. Observe donor cells under 40× objective and aspirate donor cells into injection pipette. As the cells are picked up into the pipette, they should

be aspirated back and forth several times through the opening of the needle.

9. The cell should be distorted by the pipette when it is picked up. If the pipette is too large the cell will not be broken, if it is too small, the nucleus will be damaged.
10. Collect no more than 5–10 cells at a time.

4.4.5 Nuclear Transfer

1. Once the cells (nuclei) have been collected, move the injection pipette to the drop of medium containing the oocytes.
2. See Fig. 4.3.9 for photographic depiction of NT.
3. Immobilize the first oocyte to be injected on the holding pipette.
4. Drill through zona.
5. Expel zona "plug" into space between zona and oocyte.
6. Bring the injection pipette almost to the opposite side of the enucleated oocyte, close to the holding pipette, again making a deep furrow in the egg (like pushing one's finger into a balloon).
7. Engage a single pulse of the piezo to break the oocyte membrane, suck in a very small amount of egg cytoplasm and then immediately eject it together with the donor nucleus.
8. Leave as little PVP behind as possible.
9. Rapidly withdraw the needle from the oocyte while aspirating at the cytoplasmic membrane at the right end of the furrow.
10. By simultaneously withdrawing the needle and aspirating, it should be possible to close the hole left behind by NT. This "hole removal" technique will greatly reduce the number of oocytes that lyse during NT.
11. Each batch of enucleated oocytes should be reconstructed within 30 min or less and then returned back to KSOM at 37 °C.
12. When returning the NT embryos to the incubator, wash them three times through KSOM, and place into a fresh drop of KSOM.
13. If the donor nucleus is exposed to normal enucleation medium it will becomes sticky immediately and should be discarded.
14. 90–95% of reconstructed eggs can survive this procedure.

4.4.6 Oocyte Activation and Long-Term Culture

1. Allow the reconstructed eggs to rest in KSOM, 37 °C, for 1–3 h.
2. Pick up NT embryos in batches and transfer them to activation medium in Petri dish 4.
3. Wash NT embryos through three drops of activation medium to remove residual KSOM.
4. Place small groups of embryos (5–10) into the small drops of activation medium.
5. Return dish to incubator and culture for 5–6 h at 37 °C.
6. Following activation, wash the embryos through KSOM medium 6–7 times to remove cytochalasin B and place embryos in KSOM (Petri dish 5).
7. Culture at 37 °C until the two-cell stage for oviduct transfer or the blastocyst stage for ES cell derivation and uterine transfer.

References

Baguisi A, Behboodi E, Melican DT, Pollock JS, Destrempes MM, Cammuso C, Williams JL, Nims SD, Porter CA, Midura P, Palacios MJ, Ayres SL, Denniston RS, Hayes ML, Ziomek CA, Meade HM, Godke RA, Gavin WG, Overstrom EW, Echelard Y (1999) Production of goats by somatic cell nuclear transfer. Nat Biotechnol 17:456–461

Bartolomei MS, Tilghman SM (1997) Genomic imprinting in mammals. Annu Rev Genet 31:493–525

Barton SC, Surani MA, Norris ML (1984) Role of paternal and maternal genomes in mouse development. Nature 311:374–376

Betts D, Bordignon V, Hill J, Winger Q, Westhusin M, Smith L, King W (2001) Reprogramming of telomerase activity and rebuilding of telomere length in cloned cattle. Proc Natl Acad Sci USA 98:1077–1082

Boiani M, Eckardt S, Scholer HR, McLaughlin KJ (2002) Oct4 distribution and level in mouse clones: consequences for pluripotency. Genes Dev 16:1209–1219

Bortvin A, Eggan K, Skaletsky H, Akutsu H, Berry DL, Yanagimachi R, Page DC, Jaenisch R (2003) Incomplete reactivation of Oct4-related genes in mouse embryos cloned from somatic nuclei. Development 130:1673–1680

Briggs R, King TJ (1952) Transplantation of living nuclei from blastula cells into enucleated frogs' eggs. Proc Natl Acad Sci USA 38:455–463

Briggs R, King TJ (1957) Changes in the nuclei of differentiating endoderm cells as revealed by nuclear transplantation. J Morphol 100:269–311

Burdon T, Stacey C, Chambers I, Nichols J, Smith A (1999) Suppression of SHIP-2 and ERK signaling promote self-renewal of mouse embryonic stem cells. Dev Biol 210:30–43

Campbell KH, McWhir J, Ritchie WA, Wilmut I (1996) Sheep cloned by nuclear transfer from a cultured cell line. Nature 380:64–66

Chatot CL, Lewis JL, Torres I, Ziomek CA (1990) Development of 1-cell embryos from different strains of mice in CZB medium. Biol Reprod 42:432–440

Chesne P, Adenot PG, Viglietta C, Baratte M, Boulanger L, Renard JP (2002) Cloned rabbits produced by nuclear transfer from adult somatic cells. Nat Biotechnol 20:366–369

Cibelli JB, Stice SL, Golueke PJ, Kane JJ, Jerry J, Blackwell C, Ponce de Leon FA, Robl JM (1998) Cloned transgenic calves produced from nonquiescent fetal fibroblasts. Science 280:1256–1258

Clarke DL, Johansson CB, Wilbertz J, Veress B, Nilsson E, Karlstom H, Lendahl U, Frisen J (2000) Generalized potential of adult neural stem cells. Science 288:1660–1663

DiBerardino MA, King TJ (1967) Development and cellular differentiation of neural nuclear-transplants of known karyotype. Dev Biol 15:102–128

Eggan K, Akutsu H, Hochedlinger K, Rideout W III, Yanagimachi R, Jaenisch R (2000) X-chromosome inactivation in cloned mouse embryos. Science 290:1578–1581

Eggan K, Akutsu H, Loring J, Jackson-Grusby L, Klemm M, Rideout WM III, Yanagimachi R, Jaenisch R (2001) Hybrid vigor, fetal overgrowth, and viability of mice derived by nuclear cloning and tetraploid embryo complementation. Proc Natl Acad Sci USA 98:6209–6214

Eggan K, Baldwin K, Tackett M, Osborne J, Gogos J, Chess A, Axel R, Jaenisch R (2004) Mice cloned from olfactory sensory neurons. Nature 428:44–49

Fulka J Jr, First NL, Moor RM (1996) Nuclear transplantation in mammals: remodelling of transplanted nuclei under the influence of maturation promoting factor. Bioessays 18:835–840

Gurdon JB, Colman A (1999) The future of cloning. Nature 402:743–746

Hochedlinger K, Jaenisch R (2002a) Monoclonal mice generated by nuclear transfer with mature B and T donor cells. Nature 415:1035–1038

Hochedlinger K, Jaenisch R (2002b) Nuclear transplantation: lessons from frogs and mice. Curr Opin Cell Biol 14:741–748

Hogan B, Beddington R, Costantini F (1994) Manipulating the mouse embryo.. Cold Spring Harbor Laboratory Press, Cold Spring Harbor, NY, pp 253–289

Humpherys D, Eggan K, Akutsu H, Friedman A, Hochedlinger K, Yanagimachi R, Lander E, Golub T, Jaenisch R (2002) Abnormal gene expression in mice cloned from ES and cumulus cell nuclei. Proc Natl Acad Sci USA 99:12889–12894

Humpherys D, Eggan K, Akutsu H, Hochedlinger K, Rideout WM III, Biniszkiewicz D, Yanagimachi R, Jaenisch R (2001) Epigenetic instability in ES cells and cloned mice. Science 293:95–97

Inoue K, Kohda T, Lee J, Ogonuki N, Mochida K, Noguchi Y, Tanemura K, Kaneko-Ishino T, Ishino F, Ogura A (2002) Faithful expression of imprinted genes in cloned mice. Science 295:297

Kafri T, Ariel M, Brandeis M, Shemer R, Urven L, McCarrey J, Cedar H, Razin A (1992) Developmental pattern of gene-specific DNA methylation in the mouse embryo and germ line. Genes Dev 6:705–714

Lanza RP, Cibelli JB, Blackwell C, Cristofalo VJ, Francis MK, Baerlocher GM, Mak J, Schertzer M, Chavez EA, Sawyer N, Lansdorp PM, West MD (2000) Extension of cell life-span and telomere length in animals cloned from senescent somatic cells. Science 288:665–669

McCreath KJ, Howcroft J, Campbell KH, Colman A, Schnieke AE, Kind AJ (2000) Production of gene-targeted sheep by nuclear transfer from cultured somatic cells. Nature 405:1066–1069

Nagy, A., Gersenstein, M., Vintersten, K., Behringer, R. (eds) (2002) Manipulating the mouse embryo. 3 edn Cold Spring Harbor Laboratory Press, Cold Spring Harbor, NY

Nagy A, Gocza E, Diaz EM, Prideaux VR, Ivanyi E, Markkula M, Rossant J (1990) Embryonic stem cells alone are able to support fetal development in the mouse. Development 110:815–821

Nagy A, Rossant J, Nagy R, Abramow-Newerly W, Roder JC (1993) Derivation of completely cell culture-derived mice from early-passage embryonic stem cells. Proc Natl Acad Sci USA 90:8424–8428

Ogonuki N, Inoue K, Yamamoto Y, Noguchi Y, Tenumura K, Suzuki O, Nakayama H, Doi K, Ohtomo Y, Satoh M, Nishida A, Ogura A (2002) Early death of mice cloned from somatic cells. Nat Genet 30:253–254

Onishi A, Iwamoto M, Akita T, Mikawa S, Takeda K, Awata T, Hanada H, Perry AC (2000) Pig cloning by microinjection of fetal fibroblast nuclei. Science 289:1188–1190

Polejaeva IA, Chen SH, Vaught TD, Page RL, Mullins J, Ball S, Dai Y, Boone J, Walker S, Ayares DL, Colman A, Campbell KH (2000) Cloned pigs produced by nuclear transfer from adult somatic cells. Nature 407:86–90

Rideout WM III, Eggan K, Jaenisch R (2001) Nuclear cloning and epigenetic reprogramming of the genome. Science 293:1093–1098

Rideout WM III, Wakayama T, Wutz A, Eggan K, Jackson-Grusby L, Dausman J, Yanagimachi R, Jaenisch R (2000) Generation of mice from wild-type and targeted ES cells by nuclear cloning. Nat Genet 24:109–110

Shin T, Kraemer D, Pryor J, Liu L, Rugila J, Howe L, Buck S, Murphy K, Lyons L, Westhusin M (2002) A cat cloned by nuclear transplantation. Nature 415:859

Tamashiro KL, Wakayama T, Akutsu H, Yamazaki Y, Lachey JL, Wortman MD, Seeley RJ, D'Alessio DA, Woods SC, Yanagimachi R, Sakai RR (2002) Cloned mice have an obese phenotype not transmitted to their offspring. Nat Med 8:262–267

Tian XC, Xu J, Yang X (2000) Normal telomere lengths found in cloned cattle. Nat Genet 26:272–273

Tsunoda Y, Kato Y (1997) Full-term development after transfer of nuclei from 4-cell and compacted morula stage embryos to enucleated oocytes in the mouse. J Exp Zool 278:250–254

Wakayama T, Perry AC, Zuccotti M, Johnson KR, Yanagimachi R (1998) Full-term development of mice from enucleated oocytes injected with cumulus cell nuclei. Nature 394:369–374

Wakayama T, Rodriguez I, Perry AC, Yanagimachi R, Mombaerts P (1999) Mice cloned from embryonic stem cells. Proc Natl Acad Sci USA 96:14984–14989

Wakayama T, Yanagimachi R (1999) Cloning of male mice from adult tail-tip cells. Nat Genet 22:127–128

Wakayama T, Yanagimachi R (2001) Mouse cloning with nucleus donor cells of different age and type. Mol Reprod Dev 58:376–383

Wilmut I, Schnieke AE, McWhir J, Kind AJ, Campbell KH (1997) Viable offspring derived from fetal and adult mammalian cells. Nature 385:810–813

Young LE, Sinclair KD, Wilmut I (1998) Large offspring syndrome in cattle and sheep. Rev Reprod 3: 155-63.

5 Large Insert Transgenesis

Shiaoching Gong, Nat Heintz

5.1 Introduction

The overexpression of genes in transgenic animals is an important tool for the study of gene function and regulation. In eukaryotic cells, the expression of a gene depends upon the presence of certain DNA sequences that regulate the rate of gene transcription. There are two classes of these transcriptional regulatory elements: enhancers and promoters. Promoters are regulatory elements present in the same chromosome as the gene whose activity they regulate and usually located immediately upstream of it. Enhancers are regulatory sequences that also have to be present in the same chromosome as the gene they activate, but they can be located up to hundreds of kilobases away, either upstream or downstream of their target gene. By themselves, neither enhancers nor promoters can achieve expression of genes. The functional expression of genes requires the combined activity of enhancers and promoters located within a certain distance from each other, on the same chromosome. A frequent goal of transgenic experiments is to express an exogenous gene (the transgene) by mimicking the expression pattern of an endogenous gene. To achieve this goal, it is necessary to include, in addition to the exogenous gene of interest, promoter and enhancer sequences required for its expression. In some cases, the sequences containing the appropriate enhancer/promoter combination can be located within a few kilobases (typically 1–6 kb) upstream of the gene of interest. However, for many other genes it is impossible to recapitulate in a transgenic animal the pattern of expression of an endogenous gene by using kilobases of upstream sequence. In addition, current methods do not provide for easy identification of enhancers based solely on sequence information. One way of circumventing this problem is to generate knock-in mice by homologous recombination in embryonic stem (ES) cells. In

Shiaoching Gong: Laboratory of Molecular Biology, Howard Hughes Medical Institute, Rockefeller University,
Box 260, 1230, NY 10021, USA, e-mail: gongs@mail.rockefeller.edu

Nat Heintz: Laboratory of Molecular Biology, Howard Hughes Medical Institute, Rockefeller University, Box 260 1230 New York 10021, USA e-mail: heintz@rockvax.rockefeller.edu

Principles and Practice
Mammalian and Avian Transgenesis – New Approaches
Shirley Pease and Carlos Lois (Eds.)

this approach, an endogenous gene is precisely replaced by an exogenous transgene of interest. Thus, expression of the exogenous transgene will be regulated by the transcriptional regulatory elements that usually direct expression of the endogenous gene. The engineered ES cells are then injected into blastocysts to generate a chimeric animal, which is then bred to achieve germline transmission. However, this strategy is expensive, time consuming, and typically requires several cycles of breeding to achieve transmission of the engineered mutation.

An alternative approach to obtain faithful expression in transgenic animals is to use bacterial artificial chromosomes (BACs). BACs are circular plasmids with an average genomic insert size of 100–300 kb, which are sufficiently large in most instances to carry an entire transcription unit and the associated regulatory regions (enhancers and promoters) required for appropriate patterns of gene expression in transgenic mice. In this approach, the endogenous gene whose expression pattern is to be recapitulated is replaced by an exogenous transgene by performing homologous recombination in bacteria. The resulting modified BAC will thus contain the transcriptional regulatory elements of the endogenous gene, directing the expression of the exogenous transgene. The engineered BAC is then injected into mouse zygotes by pronuclear injection to generate transgenic animals. There are several advantages of BAC transgenesis in comparison with the ES cell knock-in approach. First, manipulating bacteria is technically simpler, faster, and cheaper than culturing ES cells. Second, BAC transgenic animals are generated by pronuclear injection into single cell embryos and, therefore, obtaining germline transmission is not a limitation, as is observed in some ES-cell generated transgenic animals. However, BAC transgenic animals contain, in addition to the added BAC transgene, the regulatory sequences of the endogenous gene engineered to direct expression of the transgene. Therefore, it is important to realize that the BAC transgenic approach is intended only for gene addition, and that it cannot be used to introduce mutations into endogenous genes.

Traditionally, overexpression of cDNA in transgenic mice has been widely used in the study of gene function and regulation (Jaenisch 1988). However, the cDNA itself is often missing the important regulatory elements, which results in "position effects", since the level of expression is often influenced by the genomic sequences flanking the site at which the transgene is integrated. Position effects may reduce, or even completely abolish, transgene expression, whether through the action of specific silencing elements or simply because the transgene has inserted into a transcriptionally inactive region of the genome. Alternatively, flanking sequences may contain regulatory elements of nearby genes that act on the promoter of the transgene as an enhancer, which can lead to ectopic expression of the transgene. BACs and P-1 derived artificial chromosomes

(PACs) have become a widespread and powerful resource in manipulating the large genomic DNA of Escherichia coli. Because of its high clonal stability, low chimerism, and ease of DNA manipulation, BAC transgenic technology has become useful for the study of gene function and regulation (Shizuya et al. 1992; Ioannou et al. 1994; Marra et al. 1997; Kelley et al. 1999). BACs have served as the primary source of archived genomic DNA for a variety of genome mapping and sequencing projects (McPherson et al. 1999; Hoskins et al. 2000; Osoegawa et al. 2000; Tao et al. 2001; Mozo et al. 1999). BAC technology provides the opportunity to insert pieces of DNA sufficiently large, in most instances, to carry an entire transcription unit and its associated regulatory regions, which are required for dosage-dependent and correct patterns of gene expression in transgenic mice. This overcomes a major problem encountered in traditional cDNA transgenic analysis.

5.2 Highly Efficient BAC Modification Based on the R6Kγ Origin of Replication

An important advance in the utilization of BACs for functional studies came with the development of methods for precise manipulation (i.e., marker/gene insertion, deletion, point mutation, etc.) of BAC DNA (Yang et al. 1997). The ability to manipulate BAC constructs opened the door to strategies for gene expression and function studies, cell marking and isolation experiments that had previously been widely employed only in invertebrate systems (Heintz 2000). Several methods have been developed to modify BAC DNA in order to generate transgenic mice. However, a rate-limiting step in the utilization of this technology to characterize large numbers of genes has been the process with which BACs can be modified by homologous recombination in E. *coli*. In order to develop a system that is suitable for high-throughput manipulation of BAC constructs for transgenesis, a simple, highly efficient, and suitable method for high-throughput studies is available (Gong et al. 2002) and is described below.

First, this system relies on a new shuttle vector modified from PLD53 (Filutowicz and Rakowski 1998; Metcalf et al. 1996), which contains an R6Kγ DNA replication origin. Its function is absolutely dependent on expression of the π protein encoded by the pir gene for replication. In other words, it can only replicate in E. *coli* expressing the π replication protein. Since BAC host cells, DH10B, do not express this protein, this vector cannot freely replicate in these cells and should not contribute to the background we have observed using other shuttle vectors. In this

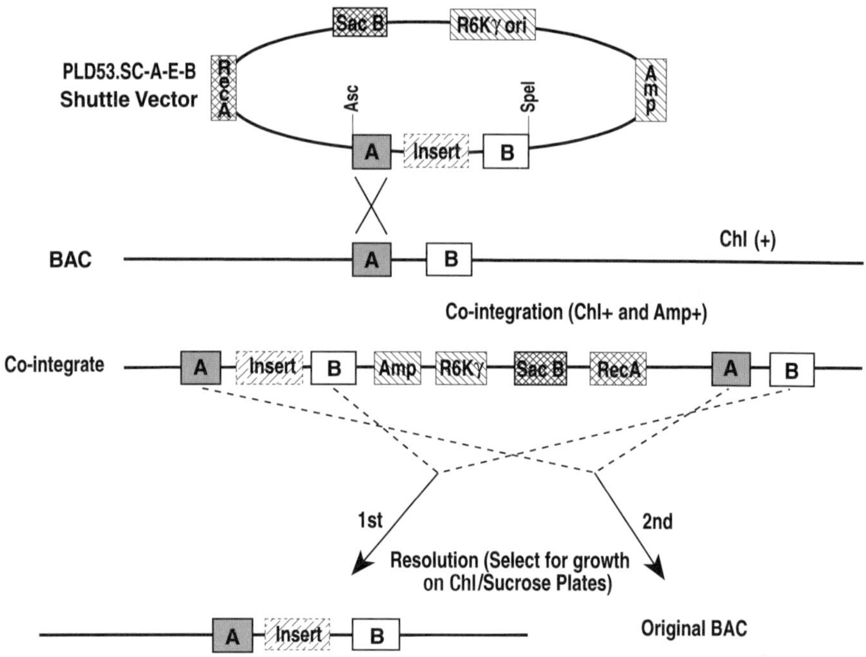

Fig. 5.1. Schematic representation of the bacterial artificial chromosome (BAC) modification

way, only bacteria containing the cointegrates are able to survive under selective conditions.

Second, since BAC libraries are maintained in a *RecA*$^{--}$ bacterial strain (DH10B), which is deficient in homologous recombination, the first step in manipulating BACs involves the restoration of competence for homologous recombination by introducing an enzyme or enzymes that can complement this deficiency. To this end, we introduce a *RecA* gene, which is useful in homologous recombination, into our shuttle vector.

Third, this shuttle vector contains a counterselection marker, the SacB gene. The *SacB* gene product, levansucrase, converts sucrose to levan, which is toxic to the host cells, thus enhancing removal of unwanted vector sequences from the manipulated BACs, by allowing selection of resolved BAC clones by culture in sucrose.

Fourth, a marker gene is also present in the shuttle vector. Analysis of the expression of this marker gene in transgenic mice facilitates exploration of individual gene expression patterns from genomic DNA.

To test this system, we PCR amplify a 500 bp homology arm (termed "A" homology arm or "A" box) from genomic DNA of the gene of interest and clone it into our shuttle vector, as shown in Fig. 5.1. We then PCR amplify another homologous arm, (termed "B" homology arm or "B" box) and clone it into the vector. The recombination cassette has two lengths of homology in arms of about 500 bp each, between which a modification (i.e., insertion, deletion, point mutation) is to be introduced. The vector is electroporated into BAC host cells, and transformants are selected in liquid LB containing chloramphenicol (Chl) and ampicillin (Amp). Since this shuttle vector contains the R6Kγ origin, it cannot replicate in the BAC host cells. Therefore, selection for both the Chl marker on the BAC and the Amp marker on the shuttle vector will allow survival of only those colonies in which cointegrates have been produced. The advantage of the R6Kγ comes from the fact that it dramatically improves the efficiency of the BAC modification procedure. The frequency of cointegration is about 50–80%.

5.3 An Approach to High Throughput Studies

5.3.1 A Precisely Modified BAC Clone for Use in the Production of Transgenic Mice (Protocol 1)

The following protocol describes a high throughput scheme for the production of precisely modified BAC clones for use in the generation of transgenic mice.

5.3.1.1 Vector Preparation

1. Transform the shuttle vector into pir2 cells, and purify the amplified DNA with a Qiagen Maxi Kit.
2. Prepare 100 μg (enough for 1,000 ligation reactions) *Asc*I/*Sma*I-digested shuttle vector by incubation overnight in appropriate amounts of enzyme. Purify the digested vector and test an aliquot in ligation to determine the background of undigested or singly digested shuttle vector. (To test digestion efficiency, ligate the cut vector alone (without insert), and transform into bacteria. The number of colonies that appear is the background, indicating undigested vector, or vector that has been digested with just one of the two enzymes.) Continue vector digestion until the background has disappeared. Aliquot and store this stock of predigested vector for use in "A box" cloning.
3. PCR amplify 300–500 bp "A box" homology regions from genomic DNA, using primers to the 3'-untranslated regions (UTRs) of the gene of interest (the 5'-primer incorporates an *Asc*I site; the 3' primer does not

incorporate any restriction sites). Digest products overnight with *Asc*I and purify digested fragments using a Qiagen QIAquick PCR purification kit.

4. Ligate the digested shuttle vector (100 ng) with each individual fragment (25 ng). Transform into pir2 cells (Invitrogen) and plate the transformed cells on Luria Bertani (LB) agar plates containing Amp (30 μg/ml).
5. Pick a few individual colonies and test for correct insertion by PCR. Insertion of the "A box" into the shuttle vector allows amplification by a combination of a 5'-end *RecA* primer and a 3'-end "A box" primer. Prepare DNA for each positive shuttle vector and confirm these clones with restriction enzymes, by comparing the digestion pattern with the vector.
6. Prepare 1 μg (enough for ten ligation reactions) *Pac*I/*Stu*I-digested shuttle vector by incubation overnight in appropriate amounts of enzyme. Purify the digested vector and test an aliquot for complete digestion as described above. Redigest it until disappearance of the background. Aliquot and store this stock of predigested vector for use in "B box" cloning.
7. PCR amplify 300–500 bp "B box" homology region from genomic DNA with a 5' primer incorporating a *Pac*I site, and a 3' primer incorporating an *Stu*I site. Digest products overnight with *Pac*I/*Stu*I and purify digested fragments using a QIAquick PCR purification kit.
8. Ligate the digested shuttle vector (100 ng) with each individual fragment (25 ng). Transform into pir2 cells (Invitrogen) and plate the transformed cells on LB+Amp (30 μg/ml) plates.
9. Pick a few individual colonies and test for the correct insertion by PCR. The insertion of "B box" into the shuttle vector allows amplification by a combination of a 5' end bovine growth hormone-PA primer and a 3' end "B box" primer. Prepare DNA for each positive shuttle vector and confirm these clones with restriction enzymes by comparing the digestion pattern with the vector.

   (a) *Note:* During this step, the A box should not contain an internal *Asc*I site. If it does, an *Mlu*I site will need to be incorporated into the 5' primer and the PCR product digested with MluI.

   (b) Invitrogen's pir2 cells are highly recommended, since this shuttle vector contains a R6kr DNA replication origin, which can replicate only in bacteria expressing the pir replication protein.

5.3.1.2
Competent Cells for Electroporation

1. Inoculate 100 ml LB with 1/1,000 volume of a fresh overnight culture.
2. Grow cells at 37 °C with vigorous shaking to OD_{600}=0.5–0.8 (in our experience this takes about 5–6 h to reach an OD of 0.7).
3. Harvest cells by centrifugation at 3,000 rpm for 10 min (Beckman J6-MI centrifuge) at −5 °C.
4. Resuspend pellets in equal volumes of 10% cold glycerol. Centrifuge as in step 3.
5. Repeat step 4 once more.
6. Take off supernatant, draining as much as possible of the residue.
7. Gently resuspend cells to a final volume of 200 μl with 10% cold glycerol.
8. Dispense 40 μl aliquots into sterile tubes and freeze.

5.3.1.3
Preparation of Cointegrates for BACs

1. Transform PLD53-modified shuttle vector (PLD53SC.A-E-B) containing the gene of interest into BAC competent cells by electroporation as follows: Thaw 40 μl BAC-containing competent cells on ice, mix with 2 μl DNA (0.5 μg/μl), and place the mixture on ice for 1 min. Transfer each sample to a cold 0.1 cm cuvette. Use a Bio-Rad Gene Pulser apparatus to carry out the electroporation (settings: 25 μF, voltage = 1.8 kV, pulse controller = 200 Ω.
2. Add 1 ml SOC (0.5% yeast extract, 2.0% tryptone, 10 mM NaCl, 2.5 mM KCl, 10 mM $MgCl_2$, 20 mM MgSO4, 20 mM glucose) to each cuvette immediately after electroporation. Resuspend the cells and transfer to a 17 × 100 nm polypropylene tube. Incubate at 37 °C for 1 h on a shaker at 225 rpm.
3. Select for transformed cells with 5 ml LB supplemented with Chl (20 μg/ml) and Amp (30 μg/ml), and incubate at 37 °C overnight. Dilute the overnight culture 1:1,000 and grow in 5 ml LB with Chl (20 μg/ml) and Amp (50 μg/ml) at 37 °C for about 8–14 h. Dilute this culture 1:5,000 and grow in selection medium at 37 °C for 8 h. Make a series of dilutions (we recommend an initial dilution of 1:5,000) and

spread aliquots (10 μl and 100 μl each) on Chl (20 μg/ml) and Amp (50 μg/ml) plates. Incubate at 37 °C overnight.

4. Pick four single colonies per plate and use each colony to inoculate 5 ml LB supplemented with Chl (20 μg/ml) and Amp (50 μg/ml). In parallel, streak these colonies onto Chl/Amp master plates and grow overnight at 37 °C. Miniprep DNA from 3 ml culture using the alkaline lysis method. Identify correct cointegrates for each clone by PCR or Southern blot.

5.3.1.4
Screening of resolved clones from the modified BACs

1. Pick two to three colonies of the cointegrate from the Chl/Amp master plates. Inoculate each colony into 1 ml LB supplemented with chloramphenicol (20 μg/ml) and incubate at 37 °C for 1 h on a shaker at 225 rpm.

2. Spread 100 μl or 10 μl of each mixture on a TE-agar plate (10 g SELECT Peptone 140, 5 g SELECT Yeast Extract and 12 g SELECT Agar, in 1 l water; adjust pH to 7.5) supplemented with Chl (20 μg/ml) and 4–4.5% sucrose. Incubate at 37 °C overnight.

3. Pick both big and small colonies and plate them on two LB agar plates supplemented with Chl (20 μg/ml). (Usually, a higher frequency of resolution comes from small colonies.) Incubate the master plate at 37 °C directly, while exposing the other plate to UV light, by placing it on a UV transilluminator (4 × 8 W, 312 nm bulbs) for 30 s. Incubate at 37 °C overnight to check for deletion of the *RecA* gene (second recombination). After resolution, colonies that have lost the excised recombination vector including the *SacB* and *RecA* genes become sensitive to UV light. Therefore, exposure to UV light may help screen out false positive clones.

4. Characterize colonies that are sensitive to UV light by PCR or by Southern blot analysis. By PCR, resolution through eukaryotic green fluorescent protein (EGFP) allows amplification either by a combination of a 5' end co-integrate (coint)-primer, which is outside the "A Box", and an EGFP AT primer, or a combination of a 3' end coint-primer outside the "A Box" and a *RecA* primer. For Southern blot analysis, inoculate each positive colony into 3 ml LB supplemented with Chl (12.5 μg/ml) only, streak the same colony onto a Chl master plate and incubate at 37 °C overnight. Miniprep DNA from those cultures using the alkaline lysis method and use the "A Box" or EGFP gene as a hybridization probe.

5. To determine whether any unexpected deletions have been generated during modification, high resolution fingerprinting of both original and

modified BACs should be carried out after unique restriction enzyme digestion. We compare these restriction fragment patterns by agarose pulsed field gel (PFG) analysis. The correlation of appropriate restriction fragments between the modified and the original BACs suggests that those modified BACs preserve the intact sequence. The only difference should be that insertion of the EGFP gene leads to a shift of one band to a higher size on the ethidium bromide-stained gel.

5.3.1.5 Troubleshooting for BAC recombination

The appearance of non-specific colonies may indicate that your competent cells or shuttle vectors are contaminated with high-copy plasmid. This contamination may result in the growth of non-specific colonies, since this plasmid can replicate freely in DH10B cells.

If no colonies appear after selection, you should check the electroporation parameters. If the shuttle vectors or competent cells contain high concentrations of salt, this may interfere with transformation.

5.3.2 Preparation of BAC DNA by Double Acetate Precipitation and CsCl Gradient Centrifugation (Protocol 2)

One of the most important factors for successful generation of transgenic mice is the quality of DNA provided for injection. There are several methods for preparing BAC DNA. We find the following protocol to be the most reliable and efficient in the generation of transgenic mice.

1. Pick a single colony of transformed bacteria from a freshly streaked Chl (20 μg/ml) agar plate; inoculate 3 ml LB medium containing Chl (20 μg/ml). Incubate at 37 °C for 8 h.

2. Transfer 0.2–1.0 ml inoculated broth (depending on cell density) into 500 ml LB containing Chl (20 μg/ml); incubate at 30 °C for 14–16 h.

3. Spin down the bacteria at 4,000 rpm for 30 min at 4 °C (J6-MI Beckman-Coulter centrifuge, JS-4.2 rotor; Beckman, Fullerton, CA). Remove all traces of supernatant by inverting the open centrifuge tube until all medium has been drained.

4. Resuspend cells in 40 ml 10 mM EDTA, pH 8.0 by pipetting, and transfer to a 250 ml bottle.

5. Add 80 ml alkaline lysis solution (0.2 *N* NaOH in 1% SDS: 2 ml 10 *N* NaOH, 10 ml 10% SDS into 88 ml dH_2O). Mix by swirling very gently and incubate for 5 min at room temperature.
6. Add 60 ml cold 2 M potassium acetate (50 ml 7.5 M potassium acetate, 23 ml glacial acetic acid and 127 ml dH_2O, stored at 4 °C). Mix by swirling very gently and incubate on ice for 5 min.
7. Spin at 11,650 rpm for 30 min at 4 °C (J-25I Beckman Avanti centrifuge, JLA-16.250 rotor).
8. Transfer supernatant into a 500 ml bottle, add 180 ml isopropanol. Mix by gently swirling. Spin at 4,000 rpm for 30 min at 4 °C (J6-MI centrifuge, JS-4.2 rotor). Decant the supernatant.
9. Dissolve the DNA pellet in 18 ml 10:50 TE (1 ml 1 M Tris, 10 ml 0.5 M EDTA into 89 ml dH_2O). Add 9 ml 7.5 M potassium acetate, mix and incubate at −70 °C for 30 min.
10. Thaw solution and centrifuge at 6,000 rpm for 10 min at 4 °C (J-25I Beckman Avanti centrifuge, JA-25.50 rotor).
11. Transfer supernatant to a new tube and add 2.5 volumes of ethanol. Spin at 11,650 rpm for 30 min at 4 °C to precipitate the DNA (J-25i Beckman Avanti centrifuge, JLA-16.250 rotor).
12. Decant supernatant and gently resuspend pellet (while still moist) in 4.4 ml TE. Dissolve, as well as possible, 10.2 g CsCl in another 4.4 ml TE. Gently mix the CsCl solution with 4.4 ml DNA solution until the CsCl has dissolved. Add 0.2 ml ethidium bromide solution (10 mg/ml dH_2O) and mix immediately. Spin at 4,000 rpm for 10 min at 4 °C to remove debris (J6-MI centrifuge, JS-4.2 rotor).
13. Remove the supernatant and load into a Beckman Quick-Seal tube (16 × 76 nm, # 342413) using a syringe and 18-gauge needle. Seal tubes carefully and place in an NVT65 rotor. (It is very important to equilibrate the tubes to be centrifuged in opposing positions: weigh them very carefully to make sure they do not differ by more than 0.05 g). Spin at 65,000 rpm overnight (>8 h) at 18 °C.
14. Carefully remove tubes from rotor, taking care not to disturb the gradient. Use a 23-gauge needle to poke a hole in the top of the tube. Using a UV light, carefully remove the band (choose bottom band if there are two) with an 18-gauge needle with the needle bevel up. Take just the band and no more (usually about 200 μl). Transfer it to a 15 ml tube and bring it up to 2 ml with TE. Extract 4–5 times with NaCl-saturated

butanol (20 ml 3 M NaCl in 100 ml butanol) until no orange color remains. (To extract, add an equal volume of NaCl-saturated butanol to the TE solution, mix gently, let mixture settle for 30 s to allow separation, remove and discard top layer.)

15. Add 1 ml H_2O to the DNA solution, and then 2.5–3.0 volumes of ethanol and mix. Place at −20 °C for 30 min. Spin solution at 11,650 rpm for 30 min at 4 °C to precipitate the DNA (J-25i Beckman Avanti centrifuge, JA-25.50 rotor). Resuspend DNA in 0.5 ml 0.3 M sodium acetate. Transfer DNA to a 1.5 ml Eppendorf tube and add 1 ml ethanol. Spin down the DNA at 14,000 rpm for 30 min at 4 °C (Eppendorf microcentrifuge model 5417R). Discard the supernatant, fill the tube with 70% ethanol and allow the tube sit at room temperature for 5 min. Spin the DNA again as in the previous step but shorten the time to 10 min. Dry the pellet at room temperature for 1 min – try to get rid of the trace amount of ethanol. Resuspend DNA gently in 20–40 μl TE. Place DNA in 37 °C incubator for 20–30 min. This procedure should yield 5–20 μg BAC DNA. BAC DNA should be stored at 4 °C. (Do not store it at −20 °C!!!).

16. Check the DNA on a PFG and determine the concentration.

17. Digest the BAC DNA with PI-*Sce*I (New England Biolabs, Beverly, MA; Cat # R0696S): mix 5–10 μl (about 100 ng) BAC DNA, 2 μl PI-SceI enzyme, 5 μl 10× buffer and dH_2O in a final volume of 50 μl. Incubate in a 37 °C incubator for 3–4 h.

18. To dialyze the DNA, start by placing 20 ml injection buffer (see Sect. 5.4.2) into a sterile Petri dish and float a 25 mm, 0.025 μm filter (Millipore, Bedford, MA; Cat. # VSWP02500) on top with the shiny side up. Load the 50 μl of digested DNA onto the top of the filter and cover the Petri dish with a lid. Allow this set-up to sit at room temperature for 4–6 h. Transfer the DNA-containing solution on top of the filter to a microcentrifuge tube and add enough injection buffer to return solution to original volume of 50 μl.

19. Check the DNA on a PFG again to confirm concentration.

    General considerations:

    (a) Never vortex cells or DNA suspensions.

    (b) Use wide bore pipette tips to avoid damaging DNA.

    (c) Use caution when handling ethidium bromide, a potent mutagen.

5.3.3 Injection of BAC DNA

To determine the quality and concentration of BAC DNA, we isolate a BAC DNA, determine its concentration using several methods, and use this DNA as a standard marker to check the concentration of other BAC DNAs. We dilute this standard marker to 1, 2, 4, 8 and 16 ng/μl and compare each individual BAC DNA with this standard marker on a PFG. We inject only DNA that shows one band, with no sign of smearing below it on the PFG. Mix a specific amount of DNA with an equal volume of 2× polyamine solution [1,000 × polyamine stock: 30 mM spermine (Sigma, tetrahydrochloride, #S-1141), 70 mM spermidine (Sigma, trihydrochloride, #S-2501] 48 h or 1 week before injection. Do not keep DNA in polyamine for longer than 1 week. If the DNA is still too sticky, dilute it in injection buffer only (containing no polyamine).

Note: Use a wide bore pipette tip for every step.

1. Inject BAC DNA (0.2–1.0 ng/μl) into 200 pronuclei of fertilized oocytes of FVB/N mice. Transfer 40 oocytes into each Swiss Webster pseudopregnant female mouse.
2. Potential transgenic founder mice that develop from the injected zygotes are screened for insertion of the transgene by PCR. The genomic DNA for this analysis is usually isolated from tail biopsies.
3. Positive founders are crossed with Swiss Webster mice and the F1 offspring are again screened by PCR for the presence of the transgene.

5.4 Tools and Materials

5.4.1 Materials

- BAC clones: Chori (bacpac.chori.org)
- Pir2 cells: Invitrogen (Carlsbad, CA)
- DNA preparation kit: Clontech (Palo Alto, CA)
- Electroporator: Bio-Rad (Richmond, CA)
- Restriction enzymes: New England Biolabs (Beverly, MA)
- Seakem LE agarose: FMC (www.fmc.com)

- Low melting point agarose: FMC
- Gel apparatus: Owl Scientific (www.owlsci.com)
- Power supply: Bio-Rad
- Pulsed field gel electrophoresis system: Bio-Rad
- PCR kit: Qiagen (Valencia, CA)
- Antibiotics: Sigma (St. Louis, MO)

5.4.2 Solutions

- 0.5 M EDTA, pH 8.0
- 1.0 M Tris, pH 8.0
- 50× TAE: Tris-base: 242 g, glacial acetic acid: 57.1 ml, QS to 1.0 l with ddH_2O.
- Buffer P1: 50 mM Tris-Cl, pH 8.0; 10 mM EDTA; 100 μg/ml RNase A
- Buffer P2: 200 mM NaOH, 1% SDS
- Buffer P3: 3.0 M potassium acetate, pH 5.5
- Injection buffer: 10 mM Tris, pH 7.5; 0.1 mM EDTA; 100 mM NaCl

References

Filutowicz M, Rakowski SA (1998) Regulatory implications of protein assemblies at the gamma origin of plasmid R6K – a review. Gene 223:195–204

Gong S, Yang XW, C Li., Heintz N (2002) Highly efficient modification of bacterial artificial chromosomes (BACs) using novel shuttle vectors containing the R6K gamma origin of replication. Genome Res 12:1992–1998

Heintz N (2000) Analysis of mammalian central nervous system gene expression and function using bacterial artificial chromosome-mediated transgenesis. Hum Mol Genet 9:937–943

Hoskins RA, Nelson CR, Berman BP, Laverty TR, George RA, Ciesiolka L, Naeemuddin M, Arenson AD, Durbin J, David RG, Tabor PE, Bailey MR, DeShazo DR, Catanese J, Mammoser A, Osoegawa K, de Jong PJ, Celniker SE, Gibbs RA, Rubin GM, Scherer SE (2000) A BAC-based physical map of the major autosomes of *Drosophila melanogaster*. Science 287:2271–2274

Ioannou PA, Amemiya CT, Garnes J, Kroisel PM, Shizuya H, Chen C, Batzer MA, de Jong PJ (1994) A new bacteriophage P1-derived vector for the propagation of large human DNA fragments. Nat Genet 6:84–89

Jaenisch R (1988) Transgenic animals. Science 240:1468–1474

Kelley JM, Field CE, Craven MB, Bocskai D, Kim UJ, Rounsley SD, Adams MD (1999) High throughput direct end sequencing of BAC clones. Nucleic Acids Res 27:1539–1546

Marra MA, Kucaba TA, Dietrich NL, Green ED, Brownstein B, Wilson RK, McDonald KM, Hillier LW, McPherson JD, Waterston RH (1997) High throughput fingerprint analysis of large-insert clones. Genome Res 7:1072–1084

McPherson JD, Marra M, Hillier, L et al (1999) A complete BAC-based physical map of the arabidopsis thaliana genome. Nat Genet 22:271–275

Metcalf et al. (1996)

Mozo et al. (1999)

Osoegawa K, Tateno M, Woon PY, Frengen E, Mammoser AG, Catanese JJ, Hayashizaki Y, de Jong PJ (2000) Bacterial artificial chromosome libraries for mouse sequencing and functional analysis. Genome Res 10:116–128

Shizuya H, Birren B, Kim UJ, Mancino V, Slepak T, Tachiiri Y, Simon, M et al (1992) Cloning and stable maintenance of 300-kilobase-pair fragments of human DNA in escherichia coli using an f-factor-based vector. Proc Natl Acad Sci USA 89:8794–8797

Tao Q, Chang YL, Wang J, Chen H, Islam-Faridi MN, Scheuring C, Wang B, Stelly DM, Zhang HB (2001) Bacterial artificial chromosome-based physical map of the rice genome constructed by restriction fingerprint analysis. Genetics 158:1711–1724

Yang XW, Model P, Heintz N (1997) Homologous recombination based modification in escherichia coli and germline transmission in transgenic mice of a bacterial artificial chromosome. Nat Biotechnol 15:859–865

6 Regional and Temporal Control of Genetic Manipulation in the Mouse

Mansuo L. Hayashi, Shigemi Hayashi

6.1 Introduction

With the deciphering of the human genome, a major challenge is to determine the function for each of the estimated 30,000 genes in the physiology and patho-physiology of various organs. Using the mouse as a model system, this has been achieved by new genetic tools that restrict gene manipulation in a regional- and time-specific manner, thus bypassing the lethality caused by the traditional gene targeting approach, which inactivates gene function in the whole body and/or during embryogenesis. The major systems for regional and temporal gene manipulation include site-specific recombinase (SSR) systems and inducible transcription systems (Branda and Dymecki 2004). Among these, the Cre-*lox*P SSR system, the Flp recombinase target (FRT) SSR system and the tetracycline (tet) regulatory systems are the most extensively used. Since the Cre-*lox*P and Flp-FRT system are quite similar in principle and methodology, in this chapter we will focus on the Cre-*lox*P, as well as the tet regulatory, system and describe their principles, applications, methodology and troubleshooting.

6.2 The Cre-*lox*P System

6.2.1 Principles

The Cre-*lox*P system utilizes Cre-mediated DNA recombination. Cre, a site-specific recombinase derived from P1 phage, binds its 34 bp recognition site *lox*P and catalyzes a recombination event between two *lox*P

Mansuo L. Hayashi: Department of Biology and the Picower Institute of Learning and Memory, Massachusetts Institute of Technology, Cambridge, MA 02139, USA, e-mail: mansuo@mit.edu

Shigemi Hayashi: Models of Disease Center, Novartis Institutes for BioMedical Research, 250 Massachusetts Ave, Cambridge, MA 02139, USA,
e-mail: shigemi.hayashi@novartis.com

Principles and Practice
Mammalian and Avian Transgenesis – New Approaches
Shirley Pease and Carlos Lois (Eds.)

sites (Fig. 6.1A). When two *lox*P sites of the same orientation are placed at the two ends of a DNA segment (i.e., the DNA segment is "floxed"), Cre-mediated recombination would excise the DNA segment between the two *lox*P sites (Fig. 6.1B). By contrast, when two *lox*P sites of opposite orientation are placed at the two ends of a DNA segment, inversion of the floxed DNA segment will occur (Fig. 6.1B). To apply this system to mice, two transgenic mouse lines usually need to be generated: one with *Cre* driven by a region- or cell type-specific promoter (referred to hereafter as "*Cre* line"), and another in which a segment of a target gene is floxed by two *lox*P sites of the same orientation (referred to hereafter as "*floxed* line"). In the doubly mutant mice, the target gene segment is deleted specifically in Cre-expressing regions and cell types (Fig. 6.1C). To completely inactivate the target gene in Cre-expressing areas, the floxed mice must be either homozygous or heterozygous (over a null allele, – that is to say one allele is deleted by traditional gene targeting, and the other allele is flanked by *lox*P sites), for the floxed gene. This system is also called the "Cre-*lox*P conditional knock-out" system. Conversely, a "Cre-*lox*P conditional activation" system has been developed in which the *floxed* line harbors a gene of interest (GOI) driven by a constitutively active promoter but silenced transcriptionally and translationally due to a floxed STOP cassette [containing an array of SV40 poly-adenylation (polyA) sequences and several translational stop codons in all reading frames] inserted between the GOI and its promoter (Fig. 6.1C).

6.2.2 Applications

Firstly, the Cre-*lox*P system has been a powerful tool for the study of gene function via conditional knock-out/activation. The recent development of mouse lines with ligand-inducible forms of Cre further offers the opportunity to temporally control Cre activity and subsequent gene manipulation at a later time following the onset of Cre expression (Garcia and Mills 2002; Hayashi and McMahon 2002). Secondly, the Cre-*lox*P system has become a popular tool for cell fate mapping studies, because Cre-mediated DNA excision is irreversible [in contrast to the tet regulatory system, which allows reversible gene manipulation (see Sect. 6.3.2)]. For example, when the *floxed* line harbors a reporter [such as β-galactosidase (β-Gal), alkaline phosphatase (AP) or green fluorescent protein (GFP), etc] following a floxed STOP cassette, the reporter is permanently expressed in Cre-expressing cells and, notably, their progeny cells (Dymecki et al. 2002). Thirdly, the Cre-*lox*P system is being applied to chromosome engineering, including inversion and translocation, for modeling of chromosomal

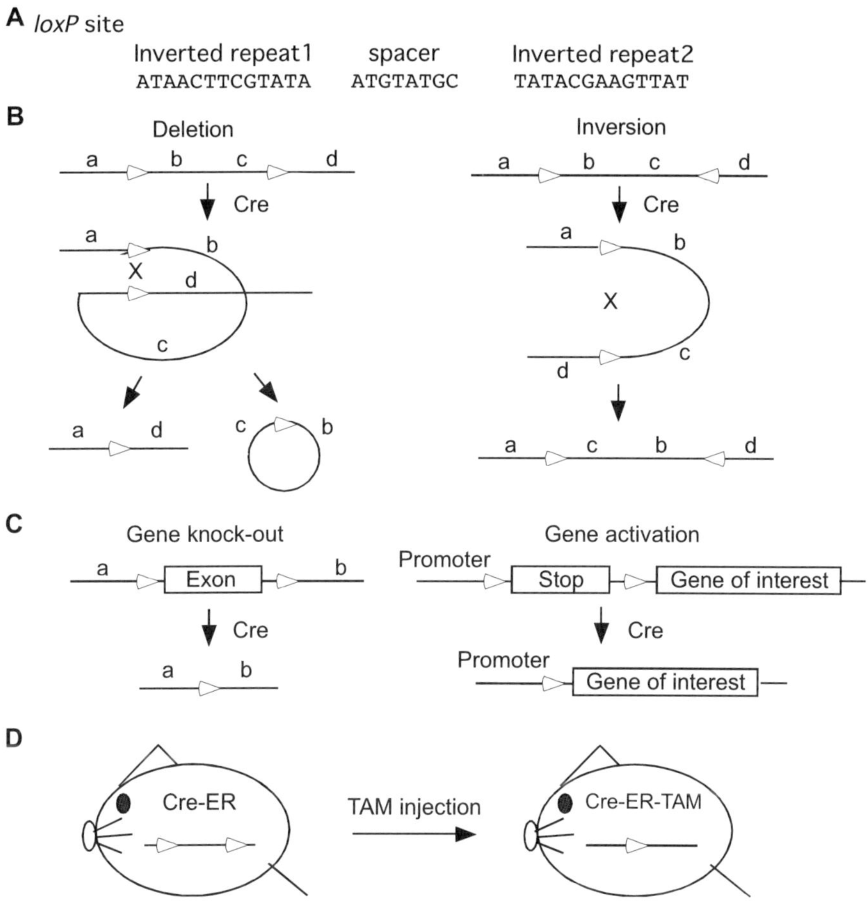

Fig. 6.1. A–D The Cre-*lox*P system. **A** The *lox*P site contains two 13 bp inverted repeats flanking an 8 bp spacer. The orientation of the spacer determines the orientation of a *lox*P site and consequently the outcome of recombination between two *lox*P sites. **B** The relative orientation of two *lox*P sites determines the outcome of a recombination event. Cre deletes the DNA segment between two directly oriented *lox*P sites (*left*) and inverts the DNA segment between two oppositely oriented *lox*P sites (*right*). **C** Schematics for the Cre-*lox*P conditional knock-out (*left*) and activation (*right*) systems. **D** In the ligand-inducible Cre-*lox*P system, binding of tamoxifen (TAM) to Cre linked to the estrogen receptor (Cre-ER) results in nuclear translocation of Cre-ER and recombination of a floxed DNA segment

rearrangements in human diseases (Branda and Dymecki 2004). Fourthly, mutated loxP sequences have been developed to elicit Cre-mediated recombination events that are independent of recombination mediated by the original (wild-type) loxP sequence (Baer and Bode 2001). This allows recombination to occur between pairs of homotypic (e.g., *lox*P/*lox*P or *lox5171*/*lox5171*) but not heterotypic (e.g., *lox*P/*lox5171*) sites, thus providing the basis for stable exchange or inversion of DNA segments (Branda and Dymecki 2004). Due to space limitations, we will describe methodology related only to the first and second applications.

6.2.3 Generating Cre lines

The general methodology of the Cre-*lox*P system has been covered by several excellent reviews (Lewandoski 2001; Ryding et al. 2001; Kuhn and Schwenk 2002). General techniques and considerations for production of transgenic/knock-out mice, mouse handling, breeding, and genotyping have been described elsewhere (Nagy et al. 2003). As for the time frame, cloning of DNA constructs takes 1–3 months, generation of *Cre* and *floxed* lines takes 1 month (in the case of transgenic approach) or 6 months (in the case of knock-in approach), and founder screening takes 3–6 months. Therefore, the analysis of doubly mutant mice usually starts at 6–12 months following the onset of the project.

A list of most, if not all, available *Cre* lines with corresponding reference and contact information is constantly updated in the Cre transgenic database on the Nagy group website (http://www.mshri.on.ca/nagy). A new *Cre* line can be generated through either a transgenic or a knock-in approach. In the transgenic approach, the DNA construct consists of the *Cre* coding sequence sandwiched between a promoter and an intron/polyA sequence. (The choice of promoter depends on which region and cell type you wish to manipulate.) The *Cre* sequence is fused to a nuclear localization signal (NLS), to localize Cre in the nucleus and maximize Cre activity. The intron/polyA sequence, e.g., the SV40 intron/polyA fragment [*Xho*I/*Bam*HI-digested product of pMSG (Pharmacia, Uppsala, Sweden)], is used to promote NLS-Cre expression. After the completion of cloning, the DNA construct is purified, linearized and injected into the pronuclei of fertilized eggs to obtain transgenic founders (Nagy et al. 2003). In some cases, a gene that is expressed in a desired region and cell type has been identified, but its promoter has not been defined. In this case, a bacterial artificial chromosome (BAC) clone that contains the genomic sequence of the putative promoter can be used. If the BAC clone also carries the gene's translational initiation codon, *NLS-Cre* is recombined into the BAC clone, replacing the translational initiation codon, and the *NLS-Cre*-containing BAC is then injected into fertilized eggs. (For BAC-related methodology, please see Chap. 4, Large Insert Transgenesis). If the BAC clone does not carry the gene's translational initiation codon, the BAC clone can be coinjected into eggs with a DNA construct carrying *NLS-Cre* driven by a minimal promoter (Nakazawa et al. 2002). When the BAC clone and the minimal promoter-*NLS-Cre* construct cointegrate into the chromosome, NLS-Cre would be expressed in the region and cell type defined by the BAC clone. To make a *Cre* line using a knock-in approach, *NLS-Cre* is recombined into a desired gene locus in embryonic stem (ES) cells, which are then injected into blastocysts to obtain chimeric founders (Nagy et al. 2003).

6.2.4 Genotyping *Cre* Lines

Cre lines are genotyped by the presence of *Cre*. We use the following primers and PCR conditions: 5'-ACCTGATGGACATGTTCAGGGATCG-3' and 5'-TCCGGTTATTCAACTTGCACCATGC-3'; 94 °C for 2 min, 30 cycles of 94 °C for 50 s, 60 °C for 1 min, 72 °C for 1 min, then 72 °C for 7 min. However, when different *Cre* lines need to be identified by genotyping, we would design one oligonucleotide in the 3' region of the promoter and the other complementary to the 5' region of *Cre*.

6.2.5 Screening *Cre* Lines

After confirmation of *Cre* by PCR and Southern blot analysis, positive founders are screened initially by their Cre expression level, first by immunohistochemistry and then by crossing with Cre-responsive reporter lines, to determine in which region and cell lineage recombination occurs. The reporter assay directly assesses the efficiency of Cre-mediated recombination, thus providing a more faithful indicator of Cre activity than Cre expression analysis. It should be cautioned, however, that the reporter assay may not predict the efficiency of the recombination event of a GOI in a specific *floxed* line, because the recombination efficiency could vary due to the chromosomal locus of the floxed allele and the distance between two loxP sites. Thus, one should determine the frequency and kinetics of target gene activation/inactivation with the specific *Cre* line and *floxed* line to be used for phenotypic analysis.

Currently available Cre-responsive reporter lines are listed in Table 2 of (Branda and Dymecki 2004). The most popular reporter lines are *R26R* and *Z/AP* mice. In R26R mice, *β-Gal*, together with an upstream floxed

STOP cassette, is knocked into a constitutively and ubiquitously active locus *Rosa 26* (Soriano 1999). In doubly mutant mice *Cre;R26R*, β-Gal is expressed in regions and cell lineages where the *STOP* cassette is deleted by Cre-mediated recombination. In *Z/AP* mice, *AP*, driven by the CAG promoter (chicken β-actin promoter coupled with cytomegalovirus enhancer), is silenced by a floxed *STOP* cassette that contains the β-*Gal* sequence (Lobe et al. 1999). In doubly mutant mice *Cre;Z/AP*, β-Gal is expressed in regions and cell lineages without Cre-mediated recombination, whereas AP is expressed in regions and cell lineages with Cre-mediated recombination. This double reporter system thus provides a convenient and reliable indicator of Cre activity. Protocols to detect Cre, β-Gal and AP are described below. When GFP and other color variant fluorescent proteins are used as reporters, the whole embryo/tissue or sections of animal/tissue can be directly examined under a fluorescence microscope.

6.2.5.1 Immunohistochemistry for Cre

For Cre staining in tissue sections, the mice are perfused with cold saline (0.9% NaCl) for 10 min, and cold 4% paraformaldehyde (PFA)/phosphate-buffered saline (PBS) for 30 min and the tissue dissected out and post-fixed in the same fixative at 4 °C overnight. On the following day, the tissue is sectioned in cold PBS at 50 μm thickness by a Vibratome. Sections can then be stored in PBS at 4 °C for up to 6 months, until use. Sections are blocked in 5% goat serum/0.2% Triton X-100/PBS at room temperature for 1.5 h, and incubated with a rabbit anti-Cre antibody (Covance, Berkeley, CA/Babco, Richmond, CA) at 1:2,000–3,000 dilution in 1% goat serum/0.1% Triton X-100/PBS, with gentle shaking at 4 °C overnight. After washing three times with PBS (each for 10 min), sections are incubated with appropriately diluted secondary antibody [e.g., Alexa 488-conjugated anti-rabbit IgG (Molecular Probes, Eugene, OR) at 1:300 dilution] in 1% goat serum/0.1% Triton X-100/PBS, at room temperature for 1.5 h. After washing three times with PBS, (each for 10 min), sections are mounted onto slides by floating in PBS, then air-dried and overlaid with coverslips under Vectashield mounting medium (Vector Laboratories, Burlinghame, CA). To localize Cre expression, sections can be double-stained with anti-Cre antibody and an antibody that is produced from a species other than rabbit.

6.2.5.2
Immunohistochemistry for β-Gal

β-Gal staining follows the same procedures as Cre staining above, except that a rabbit anti-β-gal antibody (Cappel, Aurora, OH), at 1:2,000 dilution, is used as the primary antibody.

6.2.5.3
X-Gal Staining

For whole-mount X-gal staining of embryos and tissues, small embryos (younger than E12.5) and tissues are rinsed in PBS followed by fixation with X-gal fixative (PBS containing 1% formaldehyde, 0.2% glutaraldehyde, 2 mM $MgCl_2$, 5 mM EGTA and 0.02% NP-40), with gentle shaking at 4 °C for 30–60 min. Larger embryos (E13.5 or older) or tissues are fixed in the same fixative at 4 °C for 60–90 min. Following fixation, samples are washed three times (each for 5–10 min) in PBS/0.02% NP-40 and incubated in X-gal staining buffer (PBS containing 1 mg/ml X-gal, 5 mM potassium ferrocyanide, 5 mM potassium ferricyanide, 2 mM $MgCl_2$, 0.01% sodium deoxycholate and 0.02% NP-40), at 37 °C for 30 min to 24 h. (Note that samples should be protected from light during the X-gal reaction). After completion of staining, samples are washed three times (each for 5 min) in PBS, post-fixed in PBS/4% PFA and stored in PBS/80% glycerol, at 4 °C, or photographed. It should be noted that the penetration of X-gal becomes much weaker in embryos or tissues of larger size, thus we recommend performing X-gal staining on frozen sections. Larger embryos and embryonic tissues are fixed following the above procedure. For adult tissues, mice are perfused with cold PBS/4% PFA for 15–30 min. The desired tissue is dissected out and post-fixed in the same fixative at 4 °C for 1–2 h. After fixation, samples are washed in PBS three times (each for 5–10 min), cryoprotected in PBS/30% sucrose at 4 °C overnight, incubated in OCT medium (Tissue-Tek, Miles, Eikhart, IN) at room temperature for 30 min and transferred to −20 °C overnight. Sections are cut at 10 μm thickness with a Cryostat, placed onto polylysine-coated or ProbeOn Plus slides (Fisher, Pittsburgh, PA), air dried for 30 min and then, if necessary, stored at −20 °C for 1–2 months until use. On the day of staining, sections are taken out to room temperature for 10–15 min, washed three times with X-gal wash buffer, (X-gal staining buffer without X-gal, each for 10 min), and incubated in X-gal staining buffer at 37 °C, for 30 min to 24 h (light protected). After completion of staining, sections can be counterstained with nuclear staining reagents such as Nuclear Fast Red (PolyScientific, Bay Shore, NY), rinsed in PBS, dehydrated through a graded ethanol series and xylene, and mounted with Permount (VWR, Marietta, GA). The percentage of X-gal-positive neurons out of the total number of neurons (identified by Fast Red-positive) indicates the frequency of Cre-*loxP* recombination.

6.2.5.4
X-gal and AP Double-Staining

For whole-mount double-staining, samples are fixed similarly as for the X-gal staining protocol above, except that 0.01% sodium deoxycholate is added to X-gal fixative to allow better penetration of AP substrates. After processing with X-gal staining, samples are rinsed in PBS, incubated in PBS at 70–75 °C for 30 min to remove endogenous AP, rinsed again in PBS, washed in AP buffer, (100 mM Tris-HCl pH 9.5, 100 mM NaCl and 10 mM $MgCl_2$), for 10 min and stained with BM Purple AP substrate (Boehringer Mannheim, Mannheim, Germany) at 4 °C for 0.5–24 h. After completion of staining, samples are washed extensively with PBS/0.1% Tween-20/2 mM $MgCl_2$. For double-staining of frozen sections, sections are fixed and X-gal stained and then washed in PBS three times (each for 5 min), incubated in PBS at 70–75 °C for 30 min, rinsed again in PBS, washed in AP buffer for 10 min, and overlaid with NBT/BCIP staining buffer [100 mM Tris-HCl pH 9.5, 100 mM NaCl and 50 mM $MgCl_2$, 0.01% sodium deoxycholate, 0.02% NP-40, 337 μg/ml nitroblue tetrazolium salt (NBT; Boehringer Mannheim) and 175 μg/ml 5-bromo-4-chloro-3-indolyl phosphate, toluidinium salt (BCIP; Boehringer Mannheim)] at room temperature for 10–30 min. After completion of staining, sections are rinsed in PBS, dehydrated through a graded ethanol series and mounted. The staining signal (X-gal in blue and AP in dark red) is visualized and photographed under a light microscope.

6.2.6
Generating Floxed Lines

6.2.6.1
Conditional Knock-outs

LoxP sites are usually inserted into the introns of a target gene to preserve the gene's expression. A conditional targeting replacement construct consists of a 5'-flanking region, *loxP* site 1, the intron/exon region to be deleted, *loxP* site 2, *LFNT* cassette, a 3'-flanking region and *diphtheria toxin* (*DT-A*). The LFNT cassette consists of a second pair of loxP sites, a pair of FRT sites, and pgk-*neo* and hsv-*tk* selection markers. Thus, recombined ES clones are initially selected for the presence of a *neo* marker, then transfected with Flp or Cre to remove the LFNT cassette and selected for the absence of a *tk* marker. DT-A provides an additional marker to select against non-homologous recombinants. In addition to constructing such targeting vectors via classical cloning methods, recent advancement in BAC technology provides a new and rapid means to insert loxP sites directly into ES cells (for detailed methodology, please see Chap. 4, Large Insert Transgenesis). In both cases, the resulting ES clones are injected into blastocysts, and chimeric, germline transmitting founders are confirmed by Southern blot analysis of tissue from offspring. By crossing the *floxed* line to a desired *Cre* line, the kinetics and location of target gene removal are monitored by RNA in situ hybridization and immunohistochemistry, as described below.

6.2.6.2 Conditional Activation

In a conditional activation construct, a GOI, together with an upstream floxed STOP cassette, is inserted between a constitutively active promoter and an intron/polyA sequence. The GOI can be a wild-type, constitutively active, or dominant/negative form of a gene. When the GOI is a dominant/negative form, it should be functionally potent (considering the much lower number of copies expressed in transgenic or knock-in mice than in transfected cells), and highly specific to the gene to be manipulated, in order to avoid potential quenching of its binding partners. The GOI is monitored by epitope-tagging with a myc, flag, hemagglutinin (HA), or GFP tag. The *floxed* line can be generated through either a transgenic or knock-in approach, following procedures similar to those used in the generation of *Cre* lines above. Once founders are obtained, they are screened by crossing with a desired *Cre* line to determine the level of Cre-mediated expression of the GOI. (Note that a good founder line should also possess very low levels of GOI expression in the absence of Cre). In doubly mutant mice, the level of GOI expression is determined by either RNA in situ hybridization and/or immunohistochemistry, as described below.

6.2.7 RNA in situ Hybridization

Note that all reagents and equipment must be RNase-free [e.g., RNase-free Eppendorf tubes from Marsh (division of Abgene, www.abgene.com)], and PBS and water must be pretreated with diethyl pyrocarbonate (DEPC).

6.2.7.1 Probe Preparation

^{33}P-Labeled RNA probes are prepared via in vitro transcription from a DNA template consisting of a promoter (T7, T3 or SP6) and the probe region (300–500 bases long) and terminating at the end of the probe region (so the RNA polymerase will fall off the template). The DNA template can be generated by PCR using a pair of primers, each of which includes a T7, T3 or SP6 promoter sequence plus one of the two ends of the probe region. Alternatively, the probe region can be cloned into a vector containing the T7, T3 or SP6 promoter (e.g., pSP72; Promega, Madison, WI), which is then cut at one end of the probe region by a restriction enzyme. After purification, the DNA template is subjected to in vitro transcription using a riboprobe system T7, T3 or SP6 kit (Promega) following the manufacturer's instructions. For ^{33}P-UTP, we use EasyTides uridine 5'-triphosphate-[α-^{33}P] (NEG307H from NEN, Boston, MA). To remove the DNA template, the in vitro transcription reaction mix (15 μl) is incubated with 6 μl DNase mix (including 1 μl RNasin, 4 μl tRNA at 5 μg/ml and 1 μl RQ1 DNase) at 37 °C for 15 min. The unincorporated ^{33}P-UTP is further removed by use of a spin column (e.g., ChromaSpin-30; Clontech, Palo Alto, CA). Next, to determine the amount of ^{33}P-UTP incorporation, the purified probe is diluted 1:500 in TE buffer, of which 5 μl is transferred to 4 ml water-soluble liquid scintillator (Fisher) and counted in a liquid scintillation counter, using the ^{14}C, ^{32}S and ^{33}P range. The count must be at least 3,000 cpm, and ideally around 8,000 cpm. If the count is lower than 3,000 cpm, repeat the procedure by making new DNA template and/or new probe by adding more template.

6.2.7.2
Pretreatment of Tissue Sections

Tissues should be rapidly dissected from anaesthetized mice, placed into Peel-a-way embedding molds (VWR), and covered with OCT medium (Tissue-Tek). Embedded tissues are frozen at −20 °C overnight for immediate use, or stored at −80 °C for use within 6 months. (In the latter case, tissues should be transferred back to −20 °C the night before cutting). After equilibration at −20 °C, tissues are cut at 14 μm thickness with a cryostat, and sections are mounted onto ProbeOn Plus slides (Fisher). Mounted sections can be stored at −80 °C for up to 6 months. On the day of hybridization, sections are left at room temperature for 30 min, stacked on a slide holder and processed using a Tissue-Tek slide staining set (VWR). Sections are fixed with 4% PFA/PBS for 20 min, treated with 10 μg/ml proteinase K in 10 mM Tris-HCl pH 8.0/1 mM EDTA at 37 °C for 30 min, post-fixed with 4% PFA/PBS for 15 min, rinsed with PBS for 2 min, and treated with 0.2 M HCl for 10 min. Note that too much active proteinase K can generate holes in sections. Therefore, one should determine the appropriate concentration for proteinase K for each batch purchased. After rinsing with PBS and 0.1 M triethanolamine (TEA), sections are further incubated in 0.25% acetic anhydride/0.1 M TEA, for 10 min, to avoid non-specific binding of the probe. Following dehydration through a graded ethanol series (70% to 80% to 90% to 100%) and chloroform, sections are air-dried for 30 min.

6.2.7.3
Hybridization

Hybridization buffer [50% formamide, 10 mM Tris-HCl pH 8, 200 μg/ml tRNA (RNase-free), 1X Denhardt's solution, 10% dextran sulfate, 600 mM NaCl, 0.25% SDS and 1 mM EDTA pH 8; stored at 4 °C], is preheated to 85 °C for 10 min. The probe is diluted 1:20 in this preheated hybridization buffer, then heated at 85 °C for 5 min followed by abrupt chill in icy water for over 5 min (to maintain the probe in an open conformation). The sections are then covered with the diluted probe at 150 μl per slide ($\sim 1.2 \times 10^8$ cpm) and overlaid with a piece of Parafilm precut to the size of the slide. (Bubbles should be avoided to prevent areas of non-hybridization). The slides are then placed on several layers of Whatman paper, previously wet by 50% formamide, in a box and placed within a 55 °C incubator overnight.

6.2.7.4
Washing and Exposure

After the Parafilm is removed, sections are washed at 55 °C with 5X SSC for 30 min and 50% formamide/2X SSC for 30 min. At 37 °C, sections are then washed with TNE (10 mM Tris-HCl pH 7.6, 500 mM NaCl and 1 mM EDTA) for 10 min, treated in 12.5 μg/ml RNase A/TNE for 30 min and washed with TNE for 10 min. Next, at 55 °C, sections are washed with 2X SSC twice (each for 30 min) and 0.2X SSC twice (each for 30 min), followed by dehydration through a graded ethanol series (70% to 80% to 90% to 95%) in 0.3 M ammonium acetate and 100% ethanol, and air dried. The slides are exposed to BioMax MR film (Kodak, Rochester, NY) and developed in a darkroom.

6.2.7.5
Dipping and Development of Micro-Autoradiography

To visualize hybridization signals at a higher resolution, the slides are dipped three times in NTB3 nuclear emulsion (Kodak) diluted with glycerol (1:1). Note that the entire dipping process must be performed in the dark or under red light. After immersion, the emulsion solution on the back side of the slide is wiped away with a Kimwipe paper. The slide is air dried for 1 h, transferred to a slide box with silica gel and stored at 4 °C in the dark for 4–8 weeks. To develop micro-autoradiography, the slides are incubated in D-19 solution (Sigma, St. Louis, MO) for 5 min, dipped in water several times, incubated in F-5 solution for 5 min, extensively washed by flowing water for 15 min and air dried. The hybridization signals should appear as black dots under a light microscope. To determine which cell type and subcellular compartment shows a positive hybridization signal, the sections can be further stained with Nissl and other counter-stains.

6.2.8 Tyramide Signal Amplification (TSA) for Fluorescence Immunostaining

Regular immunofluorescence staining by incubating tissue sections with primary and secondary antibodies can be carried out following the protocol described above for Cre and β-Gal staining. To enhance detection sensitivity, we utilize the TSA system (NEN), which uses horseradish peroxidase to catalyze the deposition of biotin-labeled tyramide (amplification reagent) onto tissue sections. By following the manufacturer's instructions, we find that this system shows high sensitivity and low background in signal detection. If the primary antibody is generated from mouse, we usually use the MOM kit (Vector) together with the TSA system, to avoid nonspecific staining of mouse IgG.

6.2.9 The Temporally Controllable Cre-loxP System

Ligand-inducible forms of Cre have been developed by fusing a mutant estrogen receptor (ER) ligand-binding domain to the carboxyl terminus of Cre. Cre-ER is sequestered in the cytoplasm by Hsp90 and remains inactive until it binds to synthetic estrogen agonist 4-OH-tamoxifen (4-OH-TAM; Sigma), which disrupts the interaction between Cre-ER and Hsp90. Note that TAM (Sigma), which is converted to its active derivative 4-OH-TAM in the liver, is often administered in preference to 4-OH-TAM, because TAM is more soluble in solvents like corn oil. Using various promoters to drive Cre-ER expression, several *Cre-ER* mouse lines have shown promising potential to add temporal control to the Cre-*loxP* system (Fig. 6.1D) (Garcia and Mills 2002). For instance, using the ubiquitously active CAG promoter, Cre-ER-mediated recombination was seen to be detectable in many tissues and cell types by 15 h post TAM injection, and peaked by 48 h (Hayashi and McMahon 2002). Thus, new *Cre-ER* lines that express *Cre-ER* (one of three forms: $\textit{Cre-ER}^{TM}$, $\textit{Cre-ER}^{T}$ and $\textit{Cre-ER}^{T2}$) in a desired region and cell type can be generated with protocols similar to those described in

Sect. 6.2.3. After crossing a *Cre-ER* line to a Cre-responsive reporter line (preferably a double-reporter line, like *Z/AP*), one can determine the efficiency of recombination in a desired tissue or cell type by X-gal/AP staining and/or immunohistochemistry. An optimal degree of recombination is determined by the injection of increasing doses of TAM, up to the point at which toxicity becomes apparent (e.g., 6–8 mg for a 30 g animal, and 9 mg mg for an animal of 40 g or more). TAM is dissolved in corn oil at up to 20 mg/ml at 65 °C with gentle stirring and can be stored at 4 °C for 3–4 weeks, until a white precipitate appears. When a *Cre-ER* line is crossed to a *floxed* line in conditional knock-out/activation experiments, RNA in situ hybridization and/or immunohistochemistry should be performed to determine the degree of Cre-ER-mediated recombination.

6.3 The Tetracycline Regulatory System

6.3.1 Principles

The tet regulatory system regulates gene expression at the transcriptional level via a tet-controlled transcriptional activator (tTA). tTA is a fusion product of the repressor of *Escherichia coli Tn10* tet-resistant operon and the carboxyl-terminal portion of herpes simplex virus protein VP16 that functions as a strong transcriptional activator (Fig. 6.2A) (Gossen et al. 1993). The GOI is driven by an array of tet operator (*tetO*) sequences, followed by a minimal promoter (*Pmin*). In the absence of tet, tTA binds to *tetO* and activates GOI transcription, whereas in the presence of tet, tTA is sequestered by tet from *tetO* and GOI transcription is inactivated. To apply this system to mice, two mouse lines usually need to be generated: onc with tTA driven by a region-, or cell-type-specific, promoter (hereafter referred to as "tTA regulator line"), and the other with a GOI driven by *tetO-Pmin* (hereafter referred to as "tet responsive line"). In doubly mutant mice, GOI transcription will be active – only in the region or cells defined by the promoter upstream of *tTA* – in the absence of tet but inactive in the presence of tet. Thus, this system is also called the "tet-off" system. Conversely, the "tet-on" system has been developed via the generation of a mutant form of tTA, reverse tet-controlled transactivator (rtTA), which activates gene transcription only in the presence of tet (Fig. 6.2B).

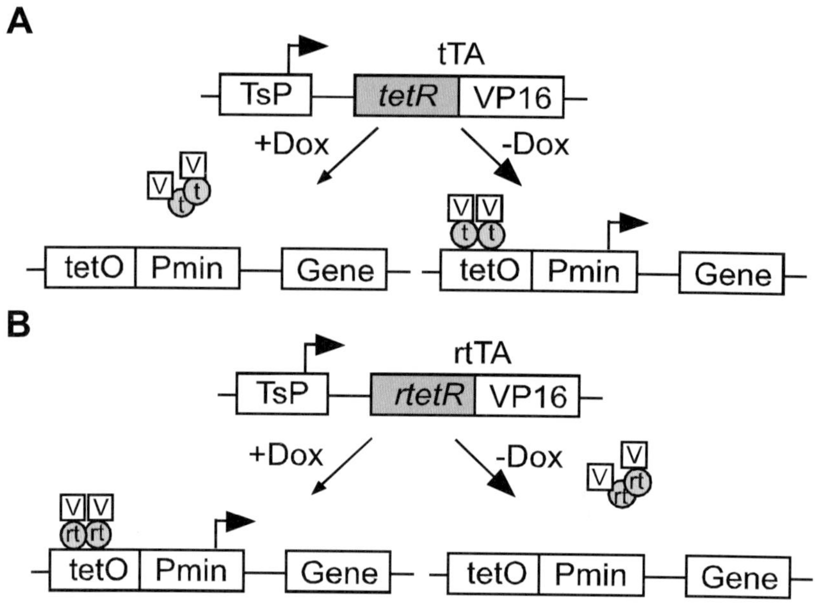

Fig. 6.2. A,B The tetracycline (tet) regulatory system. **A** *tTA*, expressed from a tissue-specific promoter (*TsP*), is a fusion protein of the repressor of *Escherichia coli* Tn10 tet-resistant operon (*tetR*) and a carboxyl-terminal portion of *VP16*. In the absence of doxycycline (dox), but not in its presence, tTA binds tet operator (*tetO*) sequences and activates transcription of a gene from a minimal promoter (*Pmin*). *B* rtTA, expressed from TsP, binds *tetO* and activates transcription of a gene from *Pmin* only in the presence of dox

6.3.2 Applications

By temporally expressing a wild-type, constitutively active, or dominant/negative form of a GOI, the tet regulatory system is used to study gene function within a desired time period, e.g., in adulthood, following the completion of development. The reversibility of gene manipulation further allows identification of critical time periods for gene function (Yamamoto et al. 2001). Thus, by temporally manipulating disease-related genes, in recent years this system has become an invaluable tool for modeling the cause and progression of human diseases in mice (Mansuy and Bujard 2000). After the appearance of disease-like symptoms, expression of these genes can be turned off to further assess the consequence of their elimination (Yamamoto et al. 2000). This is instrumental in the development of therapeutic strategies.

6.3.3
Generating tTA/rtTA Regulator Lines

The methodology of the tTA and rtTA systems is largely interchangeable. However, it should be noted that the tTA and rtTA systems differ not only in their opposite dependency on tet, or its more potent derivative doxycycline (dox), for transactivation, but also in their kinetics of transactivation/inactivation. In the tTA system, silencing of the GOI upon dox treatment was achieved in embryos after 6 h (Shin et al. 1999) and in adults after 5 days (Shimizu et al. 2000; Cui et al. 2004). Conversely, expression of a GOI upon dox withdrawal depends on the rate of dox clearance and could take from 24 h to 1 week (Kistner et al. 1996). By contrast, in the rtTA system, expression of a GOI upon dox treatment was achieved in embryos within 13 h (Shin et al. 1999) and in adults within 1 h (Hasan et al. 2001; Schonig et al. 2002). Thus, one should examine the kinetics of gene activation/inactivation with the specific combination of tTA/rtTA regulator and tet responsive lines to be used for phenotypic analysis.

In the adult brain, full rtTA-dependent activation of gene expression is achieved only by supplying dox at up to 10 mg/g food, due to the limited ability of dox to penetrate the blood brain barrier (BBB). Therefore, despite a few reports (Mansuy et al. 1998; Gogos et al. 2000), the rtTA system has not been used as extensively as the tTA system to study brain function. To overcome this, an rtTA variant with a higher dox-sensitivity (Urlinger et al. 2000) or a dox variant with higher penetration of the BBB must be developed. A general trouble-shooting guide for the tTA/rtTA system can be found at the Bujard lab website: http://www.zmbh.uni-heidelberg.de/bujard/trouble/printingver.html. Regarding the time frame for analysis, as with the Cre-*loxP* system, analysis of doubly mutant mice in the tet regulatory system usually starts at 6–12 months following the onset of the project.

A list of most, if not all, available tTA/rtTA regulator lines, with corresponding references and contact information, is available at the website http://www.zmg.uni-mainz.de/tetmouse/tet.htm. A new tTA/rtTA regulator line can be generated through either a transgenic or knock-in approach, similar to the methods described for generating *Cre* lines in Sect. 6.2.3. *tTA* can be subcloned from plasmid pUHD15–1 and *rtTA* from plasmid pUHG17–1. Both of these plasmids, together with data sheets, can be obtained from the laboratory of H. Bujard at website http://www.zmbh.uni-heidelberg.de/bujard/Homepage.html. Alternatively, similar plasmids can be purchased from Clontech.

6.3.4 Genotyping tTA/rtTA Regulator Lines

The tTA/rtTA regulator lines are genotyped by identifying the presence of tTA/rtTA. Since rtTA differs from tTA by only four amino acids (Gossen et al. 1995), the same primers can usually be used to genotype either tTA or rtTA regulator lines. For instance, we use primers 5'ATAAGCGGGCTTTGCTCGAC3' and 5'taagaaggCTGGCTCTGCAC3' and our PCR conditions are: 94 °C for 2 min, 30 cycles of 94 °C for 50 s, 60 °C for 1 min, 72 °C for 1 min, followed by 72 °C for 7 min.

6.3.5 Screening tTA/rtTA Regulator Lines

After confirmation of *tTA/rtTA* by PCR and Southern blot analysis, positive founder animals are screened by crossing with tet responsive reporter lines. A list of the available tet responsive reporter lines can be found at the website http://www.zmg.uni-mainz.de/tetmouse/strains.htm. In doubly mutant mice, the level of reporter expression is monitored both in the absence and in the presence of dox. The Bujard laboratory routinely dissolves dox-HCl (Sigma) in 5% sucrose at a final concentration of 200 μg/ml. The water bottle should be light protected and the dox-containing water should be renewed every 3 days. Alternatively, dox-containing food pellets ("dox diet" from Bio-Serv, Frenchtown, NJ) are available at multiple concentrations (20 mg/kg–6 g/kg) and can be obtained at a specific concentration by custom order. To test tet-regulated reporter expression in doubly mutant mice, we start at 1 week following feeding with dox water at 200 μg/ml or dox diet at 200 mg/kg. Depending on the result, we then alter the feeding agenda by lengthening/shortening the feeding period and/or increasing/decreasing dox concentration. When used at these doses, doxycycline is not toxic. However, if it is necessary to give very high doses of doxycycline to control the transgene, the drug will accumulate in the tissues of the animal. Thus, it will take a long time for the animal to eliminate the accumulated drug from its body before the transgene can be regulated again. For a good tTA regulator line, reporter expression should be robust in the absence of tet and quickly turned off upon dox treatment. Conversely, for a good rtTA regulator line, reporter expression should be silent (or very low) in the absence of tet, and quickly rise and reach robust levels upon dox treatment.

The most popular reporter lines are *tetO-β-gal* and *tetO-GFP*, because they allow the identification of the region and cell type within which the transgene is turned on/off. In the *tetO-GFP* system, embryos/tissues or sections of embryo/tissue from the doubly mutant mice are directly examined under a fluorescence microscope. In the *tetO-β-gal* system, mice are subjected to X-gal staining or β-gal immunohistochemistry, following protocols described in Sect. 6.2.3.

6.3.6
Generating tet-Responsive Lines

Tet-responsive lines are usually generated by pronuclear injection of a DNA construct consisting of a GOI sandwiched between *tetO-Pmin* and an intron/polyA sequence. The GOI is monitored either by epitope-tagging or by placing it under the control of a bidirectional *tetO-Pmin*. Via tTA/rtTA and dox, the bidirectional *tetO-Pmin* co-regulates transcription of two genes in opposite directions (Yamamoto et al. 2000). This enables the simultaneous transcriptional activation/inactivation of GOI and a reporter gene, such as β-gal, luciferase or GFP. Cloning vectors for both uni- and bi-directional tet-responsive lines, with data sheets, can be obtained from the Bujard laboratory on http://www.zmbh.uni-heidelberg.de/bujard/Homepage.html, or purchased from Clontech.

6.3.7
Genotyping tet-Responsive Lines

We design one primer in the *tetO* sequence (e.g., 5'TGAAAGTCGAGCTCGGTACC3') and the other complementary to part of *Pmin* or GOI. The primers are generally 20–21 bases long with 50–55% G+C and terminate with a C or G. The PCR conditions are the same as those used for genotyping tTA/rtTA regulator lines.

6.3.8
Screening tet-Responsive Lines

Due to the variability in transgene integration site and the number of integrated copies, tet-responsive lines can exhibit various levels of basal expression (in the absence of tTA/rtTA and dox) and tTA/rtTA-dependent expression. Basal expression is tested in founder mice or their progeny in the absence of tTA/rtTA and dox, while tTA/rtTA-dependent expression is tested by crossing the founder to a desired tTA/rtTA line. In these doubly mutant mice, tet regulation is further tested by varying the dose and length of dox treatment. A good tet-responsive line should exhibit a low level of basal expression, and a robust level of expression in the presence of tTA or rtTA plus dox, which is tightly regulated by dox. To monitor GOI expression, we use RNA in situ hybridization or immunohistochemistry, both of which allow the identification of regions of gene expression. As with the bi-directional tet responsive lines, it is important to verify, by in situ hybridization of consecutive sections, or by performing double immunostaining using antibodies raised against the GOI and reporter, that the GOI and the reporter do express in the same region and in the same cell types.

In the tTA system, GOI can be kept silent by dox treatment and then turned on by dox withdrawal. It should be cautioned that turning on of gene expression of GOI following dox withdrawal may not be quick, and expression may not reach the same level as in mice that have never been subjected to dox treatment. This, in our experience, occurs particularly when dox is administered during the gestation period to the pregnant mother, presumably due to incorporation of dox into bones and other tissues and consequent alteration in chromatin structure during embryogenesis. Thus, it is important firstly to determine and then to apply the lowest dose of dox in order to shut off gene expression in one's own doubly mutant mice and with one's own lab setup.

6.4 New Directions for Regional and Temporal Gene Manipulation

Firstly, quantitative control of gene activity can be achieved by varying the TAM dose in a Cre-ER-*loxP* system, or dox dose in a tet regulatory system. With a means to identify cells that have undergone gene activation/inactivation, the properties of these cells can be investigated and compared with wild-type neighbors. This would provide an efficient way to dissect the roles of gene function through cell-cell interactions. Secondly, inducible and reversible gene inactivation can be achieved by tet-regulated expression of a GOI in mice with a null or conditional null mutation of this gene. Temporal restoration of gene function in a desired region and cell type will provide a versatile tool for determination of the critical period and area of gene function (Shin et al. 1999; Shimizu et al. 2000; Cui et al. 2004). In order to overcome the time and expense involved in intercrossing multiple lines, (e.g., tTA/rtTA regulatory line, tet responsive line, knockout line), several groups have developed an elegant strategy for the insertion of a regulatory cassette (containing a STOP cassette and *tetO-Pmin*) into the 5'-UTR region of a GOI, such that transcription of this endogenous gene becomes regulated by tTA/rtTA (Bond et al. 2000; Gross et al. 2002). Thirdly, the combination of tet regulatory and Cre-*loxP* systems will provide refined temporal control to region-specific gene manipulation, when

Cre expression is regulated by tet (Utomo et al. 1999; Schonig et al. 2002). Alternatively, when tet and Cre-ER independently regulate the expression of two genes, these genes can be sequentially activated/inactivated by administration of dox and TAM, respectively. This would be advantageous in the modeling of human diseases such as cancer, where mutation in several genes occurs sequentially.

References

Baer A, Bode J (2001) Coping with kinetic and thermodynamic barriers: RMCE, an efficient strategy for the targeted integration of transgenes. Curr Opin Biotechnol pp 473–480

Bond CT, Sprengel R, Bissonnette JM, Kaufmann WA, Pribnow D, Neelands T, Storck T, Baetscher M, Jerecic J, Maylie, J et al (2000) Respiration and parturition affected by conditional overexpression of the Ca^{2+}-activated K^{+} channel subunit, SK3. Science 289:1942–1946

Branda CS, Dymecki SM (2004) Talking about a revolution: the impact of site-specific recombinases on genetic analyses in mice. Dev Cell 6:7–28

Cui Z, Wang H, Tan Y, Zaia KA, Zhang S, Tsien JZ (2004) Inducible and reversible NR1 knockout reveals crucial role of the NMDA receptor in preserving remote memories in the brain. Neuron 41:781–793

Dymecki SM, Rodriguez CI, Awatramani RB (2002) Switching on lineage tracers using site-specific recombination. Humana, Totowa, NJ

Garcia EL, Mills AA (2002) Getting around lethality with inducible Cre-mediated excision. Semin Cell Dev Biol 13:151–158

Gogos JA, Osborne J, Nemes A, Mendelsohn M, Axel R (2000) Genetic ablation and restoration of the olfactory topographic map. Cell 103:609–620

Gossen M, Bonin AL, Bujard H (1993) Control of gene activity in higher eukaryotic cells by prokaryotic regulatory elements. Trends Biochem Sci 18:471–475

Gossen M, Freundlieb S, Bender G, Muller G, Hillen W, Bujard H (1995) Transcriptional activation by tetracyclines in mammalian cells. Science 268:1766–1769

Gross C, Zhuang X, Stark K, Ramboz S, Oosting R, Kirby L, Santarelli L, Beck S, Hen R (2002) Serotonin 1A receptor acts during development to establish normal anxiety-like behaviour in the adult. Nature 416:396–400

Hasan MT, Schonig K, Berger S, Graewe W, Bujard H (2001) Long-term, noninvasive imaging of regulated gene expression in living mice. Genesis 29:116–122

Hayashi S, McMahon AP (2002) Efficient recombination in diverse tissues by a tamoxifen-inducible form of Cre: a tool for temporally regulated gene activation/inactivation in the mouse. Dev Biol 244:305–318

Kistner A, Gossen M, Zimmermann F, Jerecic J, Ullmer C, Lubbert H, Bujard H (1996) Doxycycline-mediated quantitative and tissue-specific control of gene expression in transgenic mice. Proc Natl Acad Sci USA 93:10933–10938

Kuhn R, Schwenk F (2002) Conditional knockout mice. Humana, Totowa, NJ

Lewandoski M (2001) Conditional control of gene expression in the mouse. Nat Rev Genet 2:743–755

Lobe CG, Koop KE, Kreppner W, Lomeli H, Gertsenstein M, Nagy A (1999) Z/AP, a double reporter for cre-mediated recombination. Dev Biol 208:281–292

Mansuy IM, Mayford M, Jacob B, Kandel ER, Bach ME (1998) Restricted and regulated overexpression reveals calcineurin as a key component in the transition from short-term to long-term memory. Cell 92:39–49

Mansuy IM, Bujard H (2000) Tetracycline-regulated gene expression in the brain. Curr Opin Neurobiol 10:593–596

Nagy A, Gertsenstein M, Vintersten K, Behringer R (2003) Manipulating the mouse embryo: a laboratory manual. 3 edn Cold Spring Harbor Laboratory Press, Cold Spring Harbor, NY

Nakazawa K, Quirk MC, Chitwood RA, Watanabe M, Yeckel MF, Sun LD, Kato A, Carr CA, Johnston D, Wilson MA, Tonegawa S (2002) Requirement for hippocampal CA3 NMDA receptors in associative memory recall. Science 297:211–218

Ryding AD, Sharp MG, Mullins JJ (2001) Conditional transgenic technologies. J Endocrinol 171:1–14

Schonig K, Schwenk F, Rajewsky K, Bujard H . (2002) Stringent doxycycline dependent control of CRE recombinase in vivo. Nucleic Acids Res 30:e134

Shimizu E, Tang YP, Rampon C, Tsien JZ (2000) NMDA receptor-dependent synaptic reinforcement as a crucial process for memory consolidation. Science 290:1170–1174

Shin MK, Levorse JM, Ingram RS, Tilghman SM (1999) The temporal requirement for endothelin receptor-B signalling during neural crest development. Nature 402:496–501

Soriano P (1999) Generalized lacZ expression with the ROSA26 Cre reporter strain. Nat Genet 21:70–71

Urlinger S, Baron U, Thellmann M, Hasan MT, Bujard H, Hillen W (2000) Exploring the sequence space for tetracycline-dependent transcriptional activators: novel mutations yield expanded range and sensitivity. Proc Natl Acad Sci USA 97:7963–7968

Utomo AR, Nikitin AY, Lee WH (1999) Temporal, spatial, and cell type-specific control of Cre-mediated DNA recombination in transgenic mice. Nat Biotechnol 17:1091–1096

Yamamoto A, Hen R, Dauer WT (2001) The ons and offs of inducible transgenic technology: a review. Neurobiol Dis 8:923–932

Yamamoto A, Lucas JJ, Hen R (2000) Reversal of neuropathology and motor dysfunction in a conditional model of Huntington's disease. Cell 101:57–66

7 High Resolution Gene Expression Analysis in Mice Using Genetically Inserted Reporter Genes

Niels C. Adams, Nicholas W. Gale

7.1 Introduction

Detailed analysis of the temporal and spatial patterns of expression of a gene in situ can provide a wealth of knowledge regarding the potential functional roles of a gene product. However, such detailed analysis of gene expression patterns represents a great challenge to biologists. Classical approaches such as northern blotting, RNAse protection, PCR assays and western blotting are highly reliable and sensitive means of detecting expression, but these methods completely lack spatial information indicating where the gene is expressed. Spatial information can be obtained by the use of similar approaches such as in situ hybridization and immunohistochemistry on tissue sections or whole tissue preparations. These methods, however, are limited by the availability of robust and reliable probes for the gene of interest as well as the availability of appropriately prepared tissues to probe. Even under optimal conditions, such approaches are not practical when the aim is to undertake an exhaustive screen to discover rare sites of expression in a range of ages.

The use of genetically inserted reporter genes with optimal detection characteristics, most notably the bacterial *LacZ* gene encoding β-galactosidase (β-Gal), in vivo by either targeted or transgenic insertion, allows one to screen a diverse array of tissues and developmental stages rapidly and easily, and to observe domains of expression with extremely high resolution. Coupling of reporter gene insertion with gene targeting and inactivation in mice represents an extremely powerful tool for elucidating gene function, allowing high resolution analysis of domains of expression coupled with the potential loss-of-function phenotypes of the gene of interest. This approach has been used with a great deal of success over the past 10 years in genetargeting, and considerably longer (but without the phenotypic component and with significant limitations) using

Niels C. Adams: Regeneron Pharmaceuticals, 777 Old Saw Mill River Road, Tarrytown, NY 10591, USA

Nicholas W. Gale: Regeneron Pharmaceuticals, 777 Old Saw Mill River Road, Tarrytown, NY 10591, USA, e-mail: nicholas.gale@regeneron.com

Principles and Practice
Mammalian and Avian Transgenesis – New Approaches
Shirley Pease and Carlos Lois (Eds.)

randomly integrated constructs harboring the reporter gene downstream of the presumed promoter of a gene of interest. We favor the approach of utilizing a rational and standardized construct design that allows a chosen site of integration to be targeted. We will focus our discussion on the use of this approach.

The success of gene expression analysis using reporter genes in mice is highly dependent on a well designed targeting vector. Such designs rely greatly on the ability to place the reporter gene in a precise location within an endogenous locus. Until recently, the construction of such precisely designed vectors was hampered by the limitations of classical molecular biology techniques. However, the advent of the use of bacterial homologous recombination in generating targeting vectors has largely eliminated these difficulties and now allows the execution of practically any vector design with nucleotide precision. We have developed a technology we call "VelociGene" to achieve such goals with incredible success in a very large number of gene targeting projects. The reader is referred to our description of this method and Fig. 7.1 for further details (see (Valenzuela et al. 2003)).

Despite having the ability to minimize unwanted changes that may adversely affect reporter function by manipulating the genome with great precision, factors such as RNA stability and transcriptional and translational controls can also have significant effects on the normal expression of the gene of interest. Engineering the gene of interest by inserting a reporter and deleting and/or adding controlling elements (including the introduction of a non endogenous polyadenylation signal) might result in the reporter gene not being subject to all of the normal levels of control. Therefore, despite the clear advantages of using reporter genes in mice, it is important to bear in mind the parameters of the technology. Thus, it is appropriate to view a reporter gene in mice as most accurately reflecting promoter activity of the gene of interest rather than an ultimate indication of the precise levels of expression of the gene being targeted.

This chapter incorporates the extensive experience accrued from designing and analyzing hundreds of mouse gene-targeted models, in which a variety of reporter genes have been introduced in order to understand the normal or pathological pattern of expression of the gene of interest. This chapter should serve as a guide to users of transgenic and gene targeting technology in the selection of appropriate reporter genes for expression analysis; to provide some information regarding the design of optimal reporter gene constructs; and to provide detailed methods for the optimal detection of reporter genes in situ.

Table 7.1. Comparison of reporter genes with respect to their key features. Key features in performing reporter gene analysis include the ability to perform analysis in thick tissue specimens, in which cases the reagents must be able to penetrate into such tissues, and preferably provide single-cell resolution. In addition, reporter genes vary in their sensitivity of detection, their ability to allow one to count the cells and to resolve cell shapes and processes in cells that express the reporter, and in the ability to visualize the reporter gene in living specimens. Finally the overall ease of use of the reporter is a key feature to consider when choosing a reporter gene

	Fluorescent proteins	PLAP	LacZ	Nuc-LacZ	TM-LacZ
Sensitivity	Low	Highest	High	Very high	Very high
Penetration of thick tissues	Low	Moderate	Excellent	Excellent	Excellent
Resolution in thick tissues	Low	High	High	Highest	Very high
Cell counting	No	No	No	Yes	No
Cell/process resolution	OK/good	Excellent	OK	No	Good
Viable imaging	Yes	No	No	No	No
Ease of use	Difficult	Less easy	Very easy	Very easy	Very easy

7.2 Reporter Genes

7.2.1 Selection of the Appropriate Reporter Gene

Before embarking on generating a mouse reporter knock-in, it is important to consider what the goals of the project are, because these goals can often inform one as to what the best reporter gene will be for a particular application. We will discuss the use of the best characterized reporter genes that have been used successfully in mice (see Fig. 7.1A, color plates). These include enhanced green florescent protein (eGFP), placental alkaline phosphatase (PLAP), β-galactosidase (β-Gal), and a variety of derivatives thereof. The relative advantages and disadvantages of each are discussed.

Table 7.1 briefly summarizes the various strengths of the reporter genes we have tested in our high throughput expression-screening laboratory. Subsequent tables will discuss these attributes in greater detail on a reporter-by-reporter basis. We also provide a flow diagram to guide the user in the selection of the appropriate reporter for a given scientific application (see Fig. 7.2, color plates).

7.2.1.1 Fluorescent Proteins

An ever-expanding range of fluorescent proteins is available for potential use as expression reporters (Table 7.2). One of the most widely used is eGFP. There are a great many reasons for choosing to work with eGFP as a reporter, the principal one being the ability to visualize the reporter in the tissues of living mice. Another is that eGFP has been used extensively as a fusion protein in mice with good success (see Figs. 7.3 and 7.4, color plates). However, if these are not critical aspects of the gene targeting experiment, one should probably consider other reporter genes for their various advantages over eGFP. In contrast to enzymatic reporter genes, eGFP can be visualized directly by using appropriate illumination and filter systems. However, because eGFP is a non-endogenous gene product in mice, it is amenable to detection by immunohistochemistry using anti-GFP antiserum as an alternative means of its detection. There are acceptable antisera for use in this application, as has been nicely demonstrated, for example in the GENSAT (gene expression nervous system atlas) project ((Gong et al. 2003); NINDS Contract # N01NS02331 to The Rockefeller University (New York), NY; www.gensat.org). Direct visualization of eGFP and immunohistochemistry each have their own sets of advantages and disadvantages. For example, although using immunohistochemistry to detect eGFP does not take advantage of the simplicity of direct detection, or allow visualization in living tissue, it does permit superior fixation and preservation of tissues, which allows for long-term storage and superior tissue morphology.

Table 7.2: Issues to be considered when working with fluorescent proteins as reporter genes

Resolution in thick tissues	It is extremely difficult to resolve individual cells or their processes in tissues more than 0.5 mm thick due to the limitations of optics, and signal-to-noise issues eGFP can be used to visualize whole metastatic cancers in living mice (Chishima et al. 1997)
Penetration of thick tissues	Reagent penetration is not relevant for detecting eGFP because it is directly visualized Immunohistochemistry as an alternative means of detection of eGFP, however, is highly subject to tissue penetration issues and is generally limited in use to fairly thin preparations of tissue

	The use of immunohistochemistry allows better fixation of the tissue, resulting in the ability to store tissues and to achieve better morphology. It also opens the possibility of amplifying the signal during the process. Despite the clear utility of this type of approach for some applications, it does not take full advantage that reporter genes generally afford. For example, it does not allow the screening of large pieces of tissues, and it adds an expensive and time-consuming element to the process of screening expression patterns
Sensitivity	Compared to other expression reporters, eGFP is not a sensitive mechanism for detecting gene expression
Cell counting	eGFP is generally not the first choice of expression reporter, unless counting of living cells is necessary
Cell/process resolution	eGFP can be a useful tool for tracing neuronal connections in vivo (Nguyen et al. 2002) and the dynamic processes involved in kidney development (Watanabe and Costantini 2004). Because it is a relatively small protein it is possible that it distributes well within a cell, including its processes
Viable imaging	Fluorescent proteins like eGFP are effectively the only way to visualize gene expression in living tissues and animals Fluorescent proteins are excellent reagents if the goal is to sort cells in a FACS machine. This use becomes all the more powerful when different fluorescent reporters are used to separate different populations from the same tissue sample
Ease of use	Visualization of eGFP requires proper (expensive) microscopic equipment. This also adds an element of inconvenience to the analysis Because fluorescence is frequently diminished by fixation or storage, it requires that tissues be screened shortly after harvest. This can put a significant burden on the researcher to evaluate and document results promptly eGFP is a relatively weak fluorochrome. Therefore, few endogenous promoters will drive expression at sufficiently high levels for eGFP to be easily visible. This is compounded by strong autofluorescence of mouse tissues in the range that eGPF fluoresces. This can be addressed, but only by expensive and involved spectral imaging approaches

Other variants	We have had great success using eGFP, but have been less satisfied by the results obtained when CFP (cyan fluorescent protein) and dsRed2 (both marketed by Clontech, Palo Alto, CA) are inserted into an otherwise strongly expressing locus. CFP appears to be an order of magnitude less fluorescent than eGFP, and dsRed2 has been undetectable in a configuration that generated brightly fluorescent mice when eGFP was inserted in an identical manner. We have had success recently using JRed, a novel red fluorescent protein obtained by mutagenesis of Anthomedusae jellyfish chromoprotein (marketed by Evrogen, Moscow, Russia), which, although relatively weak in florescence compared to eGFP, has good signal to background characteristics in vivo. In addition, recent success has been reported when using DsRed T3, a new variant of DsRed in mice (Vintersten et al. 2004)

Fluorescent proteins are generally amenable to fusing with other cellular proteins. Thus, they can be readily targeted to a wide variety of subcellular compartments to aid in determining which cell type expresses the gene of interest. This can also lead to a better understanding of cellular activities within that cell. A good example is the use of a histone fusion protein (H2B-eGFP) to visualize the cell nucleus in resting cells and mitotic figures in dividing cells (Pauls et al. 2001). Another example is the use of membrane-targeted (e.g., by fusion to transmembrane domain-containing sequences or by farnesylation) fluorescent proteins to aid in the visualization of cellular morphology and cellular projections such as axons and filopodia. (e.g., tau-eGFP: (Del Punta et al. 2002)).

Given that scientists are always searching for fluorescent reporters that can be more readily visualized compared to eGFP, there is a strong drive to discover and enhance naturally occurring fluorescent proteins (e.g., (Knop et al. 2002; Sun et al. 2004; Tu et al. 2003)). Recently, an alternative, yellow fluorescent protein that is faster maturing and brighter than other currently available proteins was reported (Nagai et al. 2002), raising the possibility of more widespread uses for fluorescent proteins as reporters; however, we have not yet evaluated this protein as a reporter gene in mice.

7.2.2 Placental Alkaline Phosphatase

PLAP is a well-characterized reporter protein that is frequently used in cell-based assays. We have found it to be a reliable and sensitive reporter for revealing gene expression (see Figs. 7.5 and 7.4, color plates). The histochemical reaction product has a dark-blue/purple color that is easy to see in mice, and is readily combined with counter stains in sections. A distinct advantage of PLAP over other reporters is its high sensitivity. However, PLAP does suffer from various drawbacks as outlined in Table 7.3.

Table 7.3: Issues to be considered when working with PLAP as a reporter gene

Resolution in thick tissues	The reaction product generated by PLAP is stable and allows tissues to be cleared to improve the visibility of stain. We regularly use glycerol for this purpose, though an alternative superior clearing method using BABB can be used (see Sect. 7.4.10.9), but is relatively toxic. For this reason we routinely clear in 70% glycerol
Penetration of thick tissues	When intense, the reaction product can create a relatively impermeable barrier to the staining solution, prohibiting visualization of reporter gene expression below the stained surface Providing that the whole structure being stained does not contain PLAP, then the histochemical reaction penetrates tissues reliably
	As with fluorescent reporters, good commercial antibodies to PLAP are available, but have the same caveats as for detecting eGFP
Sensitivity	Assays for detecting PLAP are very sensitive
	The use of PLAP to reveal gene expression suffers from various drawbacks, the main one being that several tissues (e.g., bone, placenta, intestine) express relatively high levels of endogenous phosphatases. This requires that endogenous activity in these tissues be heat inactivated to diminish this background, degrading tissue morphology in the process. Endogenous alkaline phosphatases can also be inhibited by Levamisole; however, in our experience, this also reduces specific signal without being sufficiently effective in reducing non-specific activity, and therefore is not recommended
Cell counting	Not an optimal reporter, given that the whole cell is stained, not just the nucleus
Cell/process resolution	PLAP is excellent for revealing cell morphology and is particularly highly valued for tracing neuronal processes (see Fig. 7.5, color plates)
	PLAP can lead to a slight fuzzy appearance of the reaction product at higher magnifications
Viable imaging	Viable imaging is not possible when using PLAP
Ease of use	PLAP is a more stable enzyme than β-Gal and as such is not as sensitive to short-term storage of tissues prior to staining
	The staining procedure requires a heat inactivation step, but is otherwise simple and relatively inexpensive to complete
	Unlike β-Gal, the PLAP staining solution can discolor non-expressing tissues if the reaction solution is not changed frequently
	As with β-Gal, the staining reaction is light sensitive

7.2.3 *LacZ* (β-Gal)

The bacterial *LacZ* gene, which encodes β-Gal, has been widely used for many years as a reporter gene (Beckwith 1981). Its use is well characterized in mice (Gossler et al. 1989; Mansour et al. 1990; Kapur et al. 1991; Bianchi et al. 2002) and in many ways it is ideally suited for this task (Table 7.4; see Figs. 7.6 and 7.4, color plates). There are various straightforward and standardized protocols for revealing the presence of β-Gal. One of the great advantages of β-Gal is that its presence can be revealed by a simple and relatively cheap reaction in the presence of X-Gal (5-bromo-4-chloro-3-indolyl β-D-glucopyranoside). The reaction generates a blue product that is visually pleasing, stable and easy to detect.

Table 7.4: Issues to be considered when working with LacZ as a reporter gene

Resolution in thick tissues	LacZ is our reagent of choice
	The histochemical product is stable and is easy to see when tissues have been cleared in glycerol.
	LacZ-stained tissues can also be cleared with BABB; however, the signal will also be cleared over a period of a few days
Penetration of thick tissues	When the protocol in this chapter is used, excellent results can be achieved when staining with thick tissues
Sensitivity	β-Gal detection is moderately sensitive, and the standard protocol is sufficient for most endogenous promoters
	When increased sensitivity is required, the histochemical reaction can be modified to make it more sensitive (see Sect. 7.4.9)
Cell counting	LacZ is not ideal for counting cells as the whole cell body can end up being stained, making cells difficult to distinguish from neighboring cells
Cell/process resolution	This is frequently possible when using LacZ, particularly when very high levels of expression are achieved
	It can be improved if LacZ is fused to tau or TM, which results in β-Gal being transported to cell processes such as axons via microtubule binding (e.g., (Feinstein et al. 2004)) or to the cell membrane in the case of TM
Viable imaging	LacZ cannot be visualized in living animals
	Other reagents [e.g. FDG (fluorescein digalactoside) from Molecular Probes, Eugene, OR] fluoresce when cleaved by β-Gal and can be used in live β-Gal-expressing cells that can be sorted in a FACS machine, though loading cells with the FDG substrate and retaining the resultant signal in the expressing cells can be challenging
Ease of use	Very easy
	The histochemical reaction is light sensitive

	One of the potential drawbacks to visualizing β-Gal with X-Gal is that one is relying on an enzymatic reaction (cleaving galactose from X-Gal). This means that care should be taken to avoid any steps that would result in protein degradation, including extended fixation and delays in processing, both of which β-Gal is sensitive to If degradation cannot be avoided, commercially available antibodies can also be used to detect the presence of β-Gal (as described above for eGFP in Sect. 7.2.1.1)
Other variants	As alluded to above, variants of LacZ created by fusing it to other protein domains are available. See Table 7.5 for further details regarding variants

This X-Gal-based assay has been routinely used to screen for expression in young embryos and in histological sections. We have adapted it for high throughput screening of gene expression in whole-mount adult tissues, as described in the protocols section. This novel use of the assay in whole adult tissues allows us to exhaustively screen for the presence of reporter gene expression throughout the entire adult mouse with a high degree of spatial resolution without having to resort to cutting serial sections through extensive sets of tissues. With practice, perfusion and tissue preparation can take as little as 20–30 min per mouse. For most applications, LacZ is the reporter gene of choice due to its ease of use and its ability to be detected with high resolution, even in thick tissue preparations.

7.2.3.1 TM-LacZ (Transmembrane LacZ), NL-LacZ (Nuclear localized LacZ) and other LacZ fusions

As with eGFP, it is possible to fuse LacZ with other protein constituents to cause LacZ to be transported to specific compartments within the cell. This approach can be considered, for example, in applications in which it would be advantageous to be able to visualize the membrane surface and/or cellular processes of the expressing cells (TM-LacZ or Tau LacZ) or if it is desirable to carefully quantify the numbers of expressing cells (NL-LacZ, see Fig. 7.4 for examples) in the genetically altered mice. When targeting genes with a signal peptide (i.e., secreted and transmembrane-domain-containing genes), it can be highly advantageous to retain the endogenous signal peptide (frequently encoded by exon 1 in such genes), and substitute a TM-LacZ reporter in downstream exons to create a fusion with the endogenous signal sequence. After postranslational processing, this TM-β-Gal protein will be deposited below the cell surface. Targeting the gene in this way allows one to avoid targeting potential regulatory sequences in the vicinity of the initiating codon of the gene of interest. Because this simplifies the design of a targeting vector, avoids the pitfalls of deleting potential regulatory sequences, and retains all of the advantages of using LacZ as a preferred reporter gene, this is our selection of choice when targeting secreted and transmembrane-domain-containing genes (see Fig. 7.1B, color plates). Table 7.5 gives examples of when the use of different LacZ fusions might be advantageous.

Table 7.5: Relative advantages to using LacZ fusions

NL-LacZ	It is frequently difficult to resolve individual cells because the LacZ protein diffuses to fill much of the cell. However, by fusing LacZ to a nuclear localizing signal, the LacZ product is sequestered to the nucleus, thus overcoming this problem Because it is discreetly localized in the nucleus, NL-LacZ allows one to readily count and quantify numbers of expressing cells NL-LacZ is generally more easily detected compared to normal LacZ, possibly due to the concentration effect of the β-Gal enzymatic activity within the confines of the nucleus, as well as the fact that the resulting signal is more discretely localized and concentrated Because the nucleus of the expressing cell is all that is visualized, one loses the important advantages of seeing cellular shapes – often a critical aid in positively identifying the cell type expressing β-Gal (see Fig. 7.4, color plates)
TM-LacZ	TM-LacZ is a good reagent for creating knock-ins into signal peptide-containing genes such as secreted and transmembrane proteins This allows insertion of the reporter gene in exons downstream of the signal peptide-encoding exons and avoids the potential complications and compromises in gene expression that may occur when a reporter and selection cassette are integrated near the promoter of a gene, or when integration inadvertently disrupts regulatory sequences in the first intron of a gene In contrast to normal LacZ, TM-LacZ fusions generally distribute over the surface of the expressing cell, which gives the staining a different quality of appearance, which sometimes aids in positively identifying the expressing cell type TM-LacZ fusions generally are more sensitive and readily detected compared to normal LacZ due to concentration of β-Gal at the cell membrane
Tau-LacZ	Tau is a protein that is associated with the cell scaffold Fusion of LacZ with tau results in β-Gal being associated with the cell's microtubules Neuronal projections and other cellular processes can be dramatically revealed by using this LacZ fusion (e.g., (Feinstein et al. 2004))

7.3 Reporter Constructs

7.3.1 Position Effects

Given the flexibility, modular design, and throughput of the VelociGene process, we have been able to test the same reporter gene in various positions within a gene of interest. In several loci, we have found that subtly changing the insertion point relative to the endogenous promoter can have significant effects on the strength of the reporter signal and fidelity of the expression pattern. Since transcription regulatory sequences often reside within the first intron, when possible, we have found it to be good practice to insert the TM-LacZ reporter cassette into the second exon to generate a transmembrane fusion reporter (see Fig. 7.1B, color plates). Because regulatory sequences can reside anywhere in a gene of interest it is valuable to compare genomic sequences from multiple species (e.g., mouse versus human) as a crude means of elucidating potential regulatory elements in non-coding regions. Where possible, one should avoid deletion of regions conserved across species. Often, however, compromises regarding the final deletion must be made to balance the desire to obliterate gene function with the desire to achieve faithful reporter gene expression. These decisions must be made on a case-by-case basis. It is also worthwhile mentioning that the proximity of the promoter and selection cassette in your targeting vector can have unwanted influences on the faithful expression of your reporter gene. Such effects should also be mitigated by placing the insertion away from the endogenous promoter, but can also be amended if the selection cassette has been flanked by loxP sites, allowing removal of the cassette by excision using Cre recombinase (see Fig. 7.1B, color plates), thus we recommend this practice.

7.3.1.1 Multiple Transgenes from the same Promoter

Although the use of internal ribosome entry site (IRES)-containing cassettes is quite widespread, we have found that the levels of the two transgenes can be rather unpredictable, with the first transgene frequently being expressed at a much higher level than the transgene following the IRES sequence. Even when a bicistronic cassette has been highly optimized, it is not likely to achieve relatively high levels of expression of either transgene, much less both of them. For these reasons, we have generally avoided the use of bicistronic approaches for revealing gene expression patterns in mice, and do not recommend them for general use.

7.4 Protocols

This section contains detailed methods for the detection of the reporter genes described above, for both whole-mount and sectioned embryos and adult tissues. When appropriate, we also describe dissection procedures in detail as well as methods to fashion helpful tools.

The recipes for standard solutions used in this section can be found at the end of the chapter.

7.4.1 Silicone Plates

There are many occasions when it is useful to immobilize tissues, either during particularly delicate or challenging dissections, or when photo-documenting results. To accomplish this, coat glass Petri dishes with a layer of silicone elastomer, and pin specimens down on this substrate using fine dissecting pins. Such a substrate is durable and clear, and allows transillumination of specimens during dissection, observation and photo-documentation. It is also sometimes useful to dissect or photograph tissues on a dark background. Adding activated charcoal to the unpolymerized silicone solution will achieve this effect. The use of silicone elastomer substrates has the added benefit of minimizing damage to delicate instruments during dissection. We generally have both clear and black plates available at all times.

Use 10 cm and 15 cm diameter glass Petri dishes coated with 5 mm Sylgard (184 Silicone Elastomer Kit; Dow Corning, Midland, MI). For plates with a black background, add enough activated charcoal (Sigma) for the unpolymerized solution to turn black. The amount will be in the order of 5% of the total volume. When the solution sets it frequently becomes lighter, so it is better to add more charcoal, than less. Best polymerization is achieved by incubating the dishes overnight at 37 °C.

7.4.2 Dissection of Embryos for Whole-mount Staining for β-Gal or PLAP visualization

7.4.2.1 General Comments

- Younger embryos (less than E12.5) generally need little dissection beyond removal from the uterus and yolk sac. Older embryos require varying degrees of sub-dissection to ensure that tissues are evenly stained.

- Since this method relies on the enzymatic properties of β-Gal or PLAP, it is important to avoid over-fixing or storing embryos prior to staining. Also, it is best to keep all reagents and tissues on wet ice throughout the dissection.
- Sacrifice pregnant dam according to local animal use policies.
- If embryos are to be genotyped, clean dissection tools before each embryo to avoid cross-contamination of the next sample.
- Use Petri dishes with a base of black silicone to aid in visualization, and so that tissues can be pinned out if the embryos need dissecting (see Sect. 7.4.1).
- All incubations take place on rockers to ensure that the tissues are gently agitated. We use a cycle rate of approximately 15–30 agitations per minute.

7.4.2.2
Materials Required

- 50 ml centrifuge tubes
- 12- or 24-well tissue culture plates (depending on the age/size of the embryo)
- 25 ml glass scintillation vials
- dissection instruments
- Tubes for collecting tissues for genotyping
- PBS (pH 8)
- Freshly made 4% paraformaldehyde (see Sect. 7.4.10)
- DNA lysis buffer (if needed for genotyping)

7.4.2.3
Dissection

1. Remove uterus and keep on ice in PBS.
2. Cut open uterus along length on side. Remove embryos, with yolk sac and placenta intact.
3. Extract embryo from yolk sac. Let embryo bleed out for a while in the PBS (hemoglobin inhibits the staining reaction and obscures positive

staining in the vasculature). Rinse with PBS. Keep embryos in PBS on ice.

4. If embryos are to be genotyped, keep yolk sac (for embryos E12.5 and younger) or cut off tail (for embryos older than E12.5) and place in DNA lysis buffer in a 96-well plate overnight at 55 °C. Embryos will have to be stained separately.
5. If embryos are E12.5 or younger, place in fixative. If embryos are E13.5 or older, dissect as described in step 6. E18.5 or older mice should be anesthetized (we use a mix of Ketamine:Xylazine 120 mg:5 mg/kg). After being anaesthetized, perfuse the embryos with 5–10 ml 4% formaldehyde. A general description of perfusion can be found in Sect. 7.4.5.2.
6. Pin down embryo, belly up, through mouth (above tongue) and at base of tail (see Fig. 7.7A).
7. Cut skin and underlying muscle straight up from umbilical cord to chin, exposing tongue and trachea. Cut down from one side of umbilical cord to leg. Cut down from other side of umbilical cord to other leg (dotted lines in Fig. 7.7A).
8. Pierce the sac surrounding the intestines at the naval to free intestines (E17 and younger).
9. Cut diaphragm at its attachment to the thorax along ribs around both sides. Pin open the sides as shown (see Fig. 7.7B).
10. Remove most of the liver, including lobe over right kidney to relieve the tight packing of internal organs by the large liver, and to allow better access of reagents to all organs.
11. Flush out with PBS.
12. Dissect the pluck (internal tissues from the tongue to the anus): gently hook the tongue and slowly pull down all of the organs in one piece, by scraping them off the spine, to below the kidneys. Make sure to include the aorta, which runs along spine. Leave pluck attached to body at the rectum (see Fig. 7.7C,D).
13. Expose the brain by cutting down the dorsal midline from the nape of the neck to the forehead. Make sure to cut open the cranium as well as the skin and pull both away from brain. Expose part of the spinal cord by making a similar dorsal midline cut (see Fig. 7.7E).
14. If embryos are not to be genotyped, all embryos from the same line can be placed in the same 50 ml tube. However, if embryos are to be geno-

typed, place individual embryos up to age E13.5 in a labeled 12-well plate or, if older, each in a 25 ml glass scintillation vial.

15. Fix in 4% paraformaldehyde with mixing at 4 °C for the times indicated in Table 7.6. Fixation times for embryos (and tissues of similar sizes) is a critical factor in obtaining good β-Gal reactions and should be determined empirically. The proper amount of fixation will stabilize the tissue and enzyme, while under- or over-fixation will lead to degradation and loss of enzymatic activity, respectively. This table provides a guide showing the fixation times that we have found to work well in our laboratory.
16. Rinse twice briefly with PBS, then wash 3x for 20 min each in large volumes of PBS at 4 °C with agitation (paraformaldehyde inhibits LacZ staining).

Table 7.6. Suggested fixation times for embryos of varying ages

Embryo age	Fix time (min) 4 °C
E8.5–E11.5	15 min
E12.5–E14.5	30 min
E15.5–E17.5	60 min

7.4.3 Whole-mount Staining of Embryos for LacZ

Examples of whole-mount stained embryos of various ages are shown in Figs. 7.8 and 7.9 (see color plates), revealing a number of interesting patterns of expression in a variety of tissues and regions of the embryo.

7.4.3.1 Materials Required

- PBS (pH 8.0)
- Rocking platform shakers (at 4 °C)
- LacZ staining solution
- 50% glycerol
- 70% glycerol
- 4% paraformaldehyde

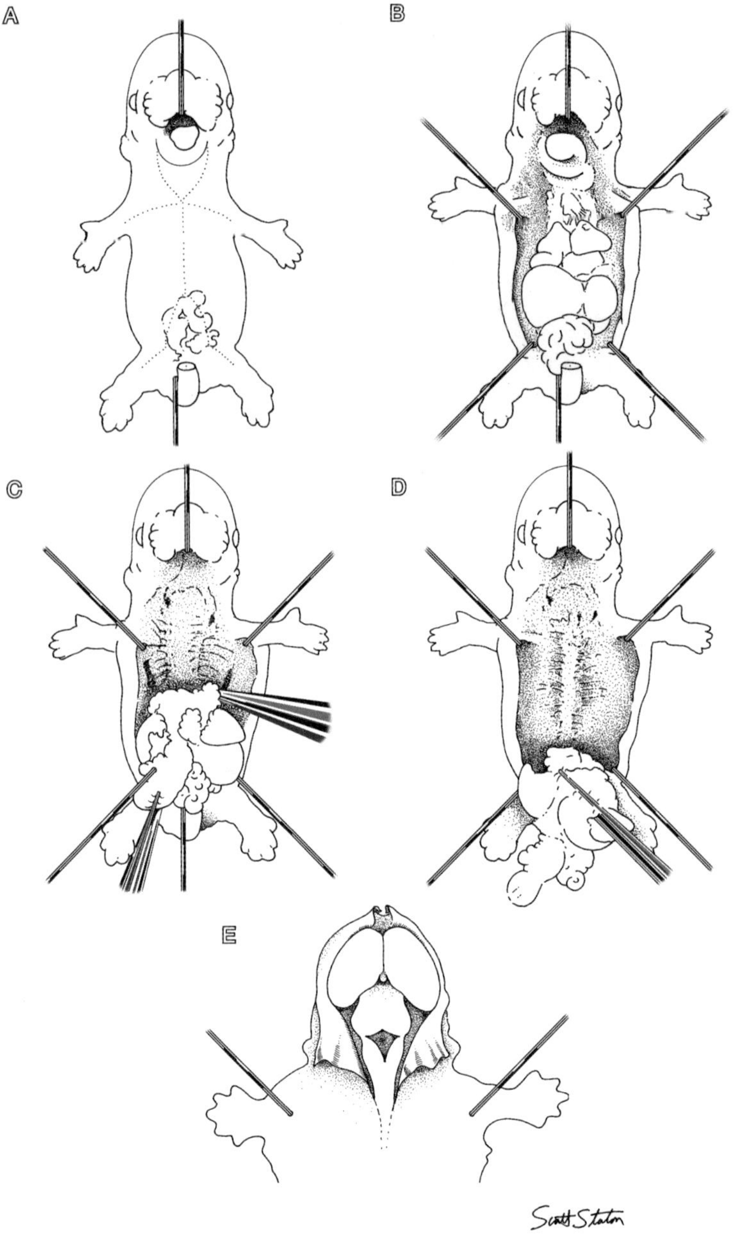

Fig. 7.7. A–E Embryo dissection guide. The illustrations shown in A–E show the manner in which embryos between embryonic day 12.5 and 17.5 are dissected to perform whole-body reporter gene expression analysis. See text Sect. 7.4.2.3 for details

7.4.3.2
Staining Method

1. Dissect and fix embryos as described in Sect. 7.4.2.3.
2. Incubate embryos in LacZ staining solution for up to 48 h in the dark at 4 °C.
3. Rinse briefly in cold PBS and then wash in cold PBS for 15 min.
4. Postfix in 4% paraformaldehyde at 4 °C overnight on the rocker.
5. Incubate in 50% glycerol on a rocker overnight at 4 °C, then clear in 70% glycerol.
6. Final storage at room temperature in 70% glycerol.

7.4.4
Whole-mount PLAP Staining of Embryos

An example of a PLAP-stained embryo can be seen in Fig. 7.5 (see color plates).

7.4.4.1
Materials Required

- Dissection instruments
- Heated water bath
- 25 ml glass scintillation vials
- 50 ml centrifuge tubes
- BCIP solution (Roche)
- NBT solution (Roche)
- PBS (pH 7.4)
- Fresh 4% paraformaldehyde in PBS
- NTMT buffer
- 50% glycerol
- 70% glycerol

7.4.4.2
Dissection and Preparation for Staining

1. Pre-heat the water bath to 75 °C.
2. Follow steps in Sect. 7.4.2.3.
3. Transfer embryos in PBS to individual 25 ml glass scintillation vials.
4. To inactivate endogenous phosphatases, heat treat the embryos in PBS for 30 min in the water bath, agitating gently every 10 min.
5. Transfer vials to ice to cool, and wash the embryos twice in PBS for 5 min each. If necessary the samples can be stored at 4 °C overnight at this point.
6. Transfer samples to 50 ml centrifuge tubes and wash three times for 20 min with cold NTMT buffer.
7. Stain the samples in NTMT buffer with 1 μl BCIP and 1 μl NBT per milliliter at 4 °C on a rocker in the dark.
8. To stop the reaction, wash embryos in PBS and then place embryos in 4% paraformaldehyde overnight.
9. Transfer the embryos to 50% glycerol overnight before storing them in 70% glycerol solution.

To minimize non-specific staining:

- Change to fresh staining solution as soon as there is a purple tinge to the solution
- Avoid exposing the staining solution to light
- Keep the staining reaction and solutions at 4 °C.

7.4.5
Dissection of Adult Tissue

Whole adult animals can be subjected to a modified version of the whole-mount analysis described above for E15.5 embryos. This technique can be highly valuable for obtaining an exhaustive survey of virtually the entire animal to determine if the reporter gene is expressed. In this assay, the organs are freed from their tight associations with other tissues, and tissues of thickness greater than approximately 2 mm are sliced at intervals to allow access of reagents. We generally employ this adult whole-mount procedure as a screening process to determine all possible sites of expression of β-Gal in the adult, and use this information to inform us as to how to proceed with high resolution analysis employing thin sections of target tissues. Examples of data from whole-mount adult tissues are shown in Figs. 7.10–7.12 (see color plates), and reveal a variety of the high quality of data that can be readily obtained from this crude technique.

7.4.5.1 Materials Required

- Perfusion pump or 50 ml syringe
- 25G syringe needles
- 125 ml amber/brown centrifuge bottles
- 50 ml centrifuge tubes
- Rocking platform shakers
- Dissection instruments
- Tubes for collecting tissues for genotyping
- PBS (pH 8)
- Freshly made 4% paraformaldehyde in PBS
- DNA lysis buffer (if genotyping mice)

7.4.5.2 Perfusion

Note: Tissue samples for northern/Southern blot and genotyping or blood samples should be collected prior to perfusion of the animal.

1. Anesthetize the mouse according to local animal use guidelines.
2. Once the mouse is clearly anesthetized (check by pinching toe and/or by eliciting a blink response), cut the skin overlying the rib cage and open the abdominal cavity by cutting the abdominal wall at the edge of the rib cage. Open the chest cavity by separating the rib cage from the diaphragm. Be careful to not damage any organs below, especially the liver, lungs and heart, which may lead to poor perfusion.
3. Cut the right atrium and perfuse through the left ventricle with cold 4% paraformaldehyde using the perfusion pump. This should take 20 min per mouse using a flow rate set at 3 ml/min.

7.4.5.3 Dissection

Note: While each organ of interest can be carefully dissected and removed from its associated tissues to be processed, the following procedure is routinely performed in our lab to make the process as rapid and thorough as possible. This method has the added advantage of keeping tissues in their normal bodily context, which can aid in identifying reporter gene expressing tissues and cells.

1. Prepare a labeled 125 ml amber bottle with 50 ml cold PBS for each animal and put it on ice. It will serve as container for tissue collection and future tissue processing.
2. Remove the skin from the surface of the rib cage. Dissect away and keep the ventral half of the rib cage containing the sternum and both cartilaginous and bony sections of rib (an example of this tissue can be seen in Fig. 7.6; see color plates).
3. Remove right ear at its base and carefully pull apart the two opposing layers of skin to expose the cartilage and muscle between them to the staining solution.
4. Remove brain from skull, then carefully slice brain with a razor along the midline. Cut brain into crude 2 mm thick slabs to allow access of reagents to various parts of the brain.
5. Open the abdomen by making two lateral incisions. Keep the freed abdominal muscle tissue separated from the overlying skin. Collect two of the abdominal mammary glands, slicing one of them in half using a razor blade.
6. Dissect all tissues from the tongue to end of the colon and all of the associated tissue and viscera. Do this by gently hooking the tongue and slowly pulling down all of the organs in one piece, by scraping/severing them from the interior surface of the dorsal body wall and spine (similar to the procedure illustrated for embryos in Fig. 7.7B–D, color plates). Make sure to include the thoracic and abdominal aorta, which runs along the spine. Also gently hook the genital area and pull up all the organs in one piece, again by scraping/severing them from the interior surface of the dorsal body wall and spine. Avoid any damage to the intestine, whose contents can foul the LacZ histochemical reaction. Remove the whole collection of tissues from the rest of the animal.
7. To allow better penetration of the staining solution, the tongue, heart, lungs, liver, spleen and kidneys should have multiple small razor cuts

along the midline. We find it most convenient to fall short of cutting completely through the organ to ensure that the tissues remain connected and can be later viewed in context following staining.

8. Cut off the remaining skull and split it along the midline with a razor blade.
9. Remove the brown fat from between the shoulder blades.
10. Keep a piece of the skin from the back.
11. Remove the spinal cord/vertebrae and associated muscle, bone and other tissues and remove surrounding tissue from the anterior portion of this to expose about 5 mm of spinal cord.
12. Remove the hind limb and pull the skin off. Slit the paw open with a razor all the way to the tip of the digit, ideally splitting the footpad in half.
13. Fix all collected tissues in 4% paraformaldehyde at 4 °C for 30 min on a rocker.
14. Rinse twice briefly with PBS.
15. Wash the tissues three times for 20 min each in PBS at 4 °C (paraformaldehyde inhibits LacZ staining).

7.4.6 Preparing Embryos and Adult Tissues for Cryo-Sectioning

Following analysis of results from the above whole-mount procedures, target tissues expressing the reporter gene can be greatly narrowed, and can then be examined with the following high resolution methods.

High resolution expression analysis using the reporter genes described in this chapter can be achieved by creating sections of target tissues of varying thickness and using varying methods. Most commonly, sections are obtained through the use of one of three common methods and devices. One method utilizes a vibrating blade microtome, which is suitable for use on soft and uniform pieces of fresh, unfrozen tissues to generate fairly thick, free-floating sections (generally ranging from 30 to 500 μm in thickness). A second method utilizes a sliding microtome, which is also suitable for soft and uniform tissues that are cut when frozen to create free floating sections of intermediate thickness (generally ranging from 40 to 100 μm). Finally, another method utilizes a cryostat, which is capable of cutting a whole range of tissues, including those containing bone, and which can create sections in the thin-to-moderately thick range (generally 5–40 μm), which are generally applied directly to microscope slides while still frozen. The method used to create sections of tissue is to some degree a matter of personal preference and/or is dependent on available sectioning equipment. We use all three techniques, depending on the goals of our experiments, but as a matter of routine we employ a cryostat to create sections. We discuss in detail only the use of the cryostat as a method of performing high resolution analysis in this chapter.

7.4.6.1
Materials Required

- Dissection equipment
- 25 ml glass scintillation vials (for embryos)
- 125 ml centrifuge bottles (for adult mice)
- OCT (TissueTek)
- Embedding molds (size appropriate to tissue to be cut)
- 2-Methylbutane (isopotane)
- Dry ice
- Lint-free tissues
- PBS (pH 8.0)
- Fresh 4% paraformaldehyde in PBS (pH 8.0)
- 15% sucrose (w/v) in PBS (pH 8.0)
- 30% sucrose (w/v) in PBS (pH 8.0)

7.4.6.2
Dissection

1. Follow step 1 through 4 of Sect. 7.4.2.3 for embryos or Sect. 7.4.5 for adult tissues.
2. Fix un-eviscerated embryos or dissected tissues from adults in 4% paraformaldehyde for 30 min with agitation at 4 °C. Do not make extra cuts in adult tissue blocks as this will make sectioning more difficult. Embryos of E15.5 and older have to be slit open along the right side of the abdomen, thorax and head to allow complete fixation of the internal organs.

3. Rinse tissues twice briefly with PBS and then wash three times for 20 min each with PBS.
4. Equilibrate in 15% sucrose until tissues sink to bottom of vial/tube (overnight) at 4 °C and subsequently in 30% sucrose overnight at 4 °C.

7.4.6.3 Embedding

1. Choose convenient mold size. The mold should be large enough to ensure that the tissues are completely submersed in OCT.
2. Remove excess sucrose carefully with a lint-free tissue.
3. Tissues should be embedded according to orientation required. We routinely cut sagittal sections of embryos. We label the outside of the mold to ensure that we can orient the block to the knife without difficulty (OCT becomes opaque when it freezes).
4. Cover tissue in OCT taking care to avoid introducing bubbles. Bubbles will mean that the tissues are not supported evenly in the frozen OCT, resulting in sections that are distorted and uneven.
5. Place mold in dry-ice-chilled 2-methylbutane until completely frozen.
6. The frozen blocks can be stored short-term on dry ice or long-term at −80 °C.

7.4.7 Cryo-sectioning

7.4.7.1 Materials Required

- Cryostat
- OCT (TissueTek)
- Acid alcohol
- Poly-L-lysine solution
- Microscope slides
- Microscope coverslips
- Coverslipping medium
- Graded alcohols for dehydrating sections prior to coverslipping
- Xylene or xylene substitute for clearing prior to coverslipping

7.4.7.2 Slide Preparation

1. Clean microscope slides with acid alcohol.
2. Place a rack of clean slides for 5 min in 1:10 diluted poly-L-lysine solution at room temperature.
3. Dry slides for 1 h at 60 °C.

An alternative to coating slides with poly-L-lysine is to use charged slides available from various manufacturers, such as SuperPlus slides from VWR (Marietta, GA). Although they provide significant time savings, we find that they do not function as well as poly-L-lysine coated slides.

Another option is to use the CryoJane Tape Transfer System developed by Instrumedics (www.instrumedics.com). This system uses a two-step process for attaching cryostat sections to the microscope slide. The first involves applying a tape to the tissue block face. The section is then cut and comes away attached to the transfer tape. Tape-backed sections are applied to a microscope slide coated with a UV-polymerizable adhesive. The adhesive is activated on the slide with a UV pulse (slides, tape and UV apparatus supplied by Instrumedics). The transfer tape is then removed and the tissue remains attached to the slide. A distinct advantage of the tape transfer system is that almost completely distortion-free frozen sections can be obtained, even from very large specimens.

We routinely use this system when sectioning whole embryos and many other tissues (especially those with bone), as the sections are free from the normal distortion encountered with typical cryostat sections, and tissues containing bone and cartilage can be sectioned without resorting to decalcification steps.

7.4.7.3 Sectioning

We regularly cut 20 μm sections to stain for the presence of β-Gal. We generally prepare a parallel set of sections cut at 10 μm to examine gross morphology through the use of standard histological stains such as Hemotoxylin and Eosin. When sections of varying thickness are collected in an alternating series, be aware of the fact that many cryostats do not cut a uniformly thick section for the first section after the cutting thickness is changed. This means that a section sometimes needs to be discarded after the cutting thickness has been changed.

Once sections have been attached to the slide, it is best to allow them to air-dry for a few minutes at room temperature before storing them at –80 °C.

7.4.7.4
LacZ Staining on Slides

Note that the histochemical reaction for β-Gal is light sensitive.

1. Retrieve the slides to be stained from −80 °C storage.
2. Let the slides to come to room temperature and dry for 30 min at room temperature to increase adhesion to the slides. Do not allow moisture to condense on slides.
3. Stain overnight at 37 °C in LacZ staining solution, followed by several rinses in PBS and post fixation in 4% paraformaldehyde for 1 h at 4 °C, followed by a mild neutral red or eosin counter stain.

7.4.7.5
Coverslipping

1. Dehydrate the sections in graded alcohols and place in Xylene substitute for no more than 5 min.
2. Coverslip the slides in a ventilated hood.

7.4.8
Whole-mount Adult LacZ Staining

It is a good idea to include both positive and negative controls (for example, from a wild type and a previously characterized transgenic animal that you are certain to have β-Gal activity) when staining, especially when working with a new project/protocol. This will help to prevent being misled by false positive staining from endogenous galactosidase activity. Another useful indicator as to whether the staining protocol worked is the intestinal contents of adult mice. However, where possible, it is always best to include a positive control specimen.

1. Incubate in LacZ staining solution at 4 °C with gentle agitation for up to 48 h. Generally, there seems little to be gained in staining past 48 h. Alternatively, shorter incubations at room temperature or 37 °C can be carried out. Although higher incubation temperatures accelerate the development of the reaction product, they have the disadvantages of increasing non-specific staining in whole-mounts (see below) and also, importantly, limiting the depth of penetration of the reaction, allowing

visualization primarily of superficial staining. For these reasons we routinely develop the stain at 4 °C.

2. Rinse in cold PBS and wash in cold PBS for 15 min.
3. Postfix in paraformaldehyde at 4 °C overnight on a rocker.
4. Clear with 50% glycerol on a rocker overnight at 4 °C.
5. Final clearing and storage in 70% glycerol at room temperature.

7.4.8.1
Common Sites of Non-Specific Staining

Non-specific staining due to endogenous galactosidase activity, most commonly encountered when evaluating β-Gal activity in whole-mounts (especially with prolonged staining times and temperatures above 4 °C and/or when staining at low pH), can be misleading. Being aware of these problematic tissues, and staining negative control tissues from age- and sex-matched specimens prepared in parallel are the best defense against being misled when finding staining in them. These tissues include nasal sinuses, lachrymal glands, salivary glands (especially strong staining in males), stomach (pyloric half), intestinal contents, mesenteric lymph nodes, male reproductive organs (testes, epididymis, vas deferens, coagulation glands). In addition, when the intestine, or its contents, stain very strongly, blue colored fluids can sometimes fill lymphatic vessels, particularly surrounding the gut and in the mesentery, and give the impression of staining in lymphatic vessels. Such false positive staining in closed structures can be identified by the ability to move or drain the apparent stain.

7.4.9
Enhancements to the Standard LacZ Staining Protocol

If the promoter driving the LacZ gene is relatively weak, it is possible to modify the staining solution and replace the ferric and ferrous cyanide with NBT (see recipe at the end of this chapter). This increases the sensitivity of the staining solution by approximately 5- to 10-fold. However, if over incubated, non-targeted tissues can appear to be stained. Our experience is that sections stained at room temperature should not be incubated for longer than 2 h. We have found this enhancement method to be very useful for staining whole embryos up to E15.5. In this case, it is not recommended that embryos be stained beyond 18 h.

We have found that this NBT-containing staining solution brings the sensitivity of the stain close to that for detecting PLAP, which is the most sensitive of reporter genes. However, we have not completed side-by-side comparisons using identical promoters. In addition, like PLAP staining, the NBT/LacZ method results in a more stable deposit, resulting in staining that is more resilient when cleared with BABB.

A comparison of the adjacent sections stained using the standard and enhanced method using NBT are shown in Fig. 7.13 (see color plates).

7.4.10 Stock Solutions

7.4.10.1 PBS

- Phosphate-buffered saline (PBS) from 10x stock solution
- Adjust to pH 8.0 with sodium hydroxide

7.4.10.2 4% Paraformaldehyde

- 4% paraformaldehyde solution in PBS, pH 8.0 (made fresh each day, kept ice cold once pH has been adjusted)

Note that paraformaldehyde is toxic, carcinogenic and teratogenic. Ensure that an effective fume hood is used along with personal protective equipment.

7.4.10.3 NTMT Buffer

- 100 mM NaCl
- 100 mM Tris pH 9.5
- 50 mM $MgCl_2$
- 0.1% Tween-20
- in distilled water

7.4.10.4 X-Gal Stock

- 4% (40 mg/ml) in dimethylformamide (DMF) in glass container, protected from light.
- Store at −20 °C

7.4.10.5 LacZ Staining Solution

- 2 mM mM $MgCl_2$
- 0.01% deoxycholic acid
- 0.02% IGEPAL CA-630
- 0.1% (1 mg/ml) X-Gal in DMF
- 5 mM potassium ferrocyanide (from 100 mM stock in amber bottle)
- 5 mM potassium ferricyanide (from 100 mM stock in amber bottle) in PBS (pH 8.0)

7.4.10.6 Alternative (more sensitive) LacZ Staining Solution

- 2 mM $MgCl_2$
- 0.01% deoxycholic acid
- 0.02% IGEPAL CA630
- 0.1% (1 mg/ml) X-Gal in DMF
- 3 μl NBT solution (Roche) per 50 ml
- in PBS (pH 8.0)

7.4.10.7 50% Glycerol

- 50% Glycerol in PBS (pH 7.4)

7.4.10.8 70% Glycerol

- 70% Glycerol in PBS (pH 7.4)
- 0.01% Sodium azide

7.4.10.9 BABB Clearing Agent

- 2 parts benzyl alcohol (v/v)
- 1 part benzyl benzoate (v/v)

Note: specimens must be dehydrated in a methanol series (25%, 50%, 75%, 95% in PBS and twice in 100% methanol) prior to equilibration with BABB. Insufficient dehydration prior to clearing in BABB will result in the tissue not being properly cleared.

7.4.10.10
DNA Lysis Buffer (for 500 ml)

- 0.1 M Tris HCl pH 7.6
- 5 mM EDTA
- 0.2% SDS
- 0.2 M sodium chloride

1. Directly prior to usage, add 100 µl Proteinase K (Roche; 1373200) per 5 ml lysis buffer.
2. Incubate samples overnight at 55 °C prior to DNA extraction and genotyping.

Acknowledgements. The authors gratefully acknowledge, Melissa Dominguez and Aris Economides for critical reading of the chapter, for help in establishing the described methods, and for invaluable discussions. We also thank all current and past members of the lab for their roles in establishing and executing the methods described. We thank Scott Staton for illustrations and Brian Ephraim for expert assistance with the figures. We are grateful to members of Regeneron's VelociGene team, lead by David Valenzuela, for the generation of thousands of test subjects that have allowed the optimization of the methods described. Finally, we owe a large debt of gratitude to the management team of Regeneron Pharmaceuticals, particularly George D. Yancopoulos, for their strong leadership and generous support.

References

Beckwith J (1981) A genetic approach to characterizing complex promoters in E. coli. Cell 23:307–308

Bianchi LM, Dinsio K, Davoli, Kand Gale NW (2002) LacZ histochemistry and immunohistochemistry reveal ephrin-B ligand expression in the inner ear. J Histochem Cytochem 50:1641–1645

Chishima T, Miyagi Y, Wang X, Yamaoka H, Shimada H, Moossa A R, Hoffman RM (1997) Cancer invasion and micrometastasis visualized in live tissue by green fluorescent protein expression. Cancer Res 57:2042–2047

Del Punta K, Puche A, Adams NC, Rodriguez I, Mombaerts P (2002) A divergent pattern of sensory axonal projections is rendered convergent by second-order neurons in the accessory olfactory bulb. Neuron 35:1057–1066

Feinstein P, Bozza T, Rodriguez I, Vassalli A, Mombaerts P (2004) Axon guidance of mouse olfactory sensory neurons by odorant receptors and the beta2 adrenergic receptor. Cell 117:833–846

Gong S, Zheng C, Doughty ML, Losos K, Didkovsky N, Schambra UB, Nowak NJ, Joyner A, Leblanc G, Hatten ME, Heintz N (2003) A gene expression atlas of the central nervous system based on bacterial artificial chromosomes. Nature 425:917–925

Gossler A, Joyner AL, Rossant J, Skarnes WC (1989) Mouse embryonic stem cells and reporter constructs to detect developmentally regulated genes. Science 244:463–465

Kapur RP, Hoyle GW, Mercer EH, Brinster RL, Palmiter RD (1991) Some neuronal cell populations express human dopamine beta-hydroxylase-lacz transgenes transiently during embryonic development. Neuron 7:717–727

Knop M, Barr F, Riedel CG, Heckel T, Reichel C (2002) Improved version of the red fluorescent protein (drFP583/DsRed/RFP). Biotechniques 33:592, 594, 596–598

Mansour SL, Thomas KR, Deng CX, Capecchi MR (1990) Introduction of a lacZ reporter gene into the mouse int-2 locus by homologous recombination. Proc Natl Acad Sci USA 87:7688–7692

Nagai T, Ibata K, Park ES, Kubota M, Mikoshiba K, Miyawaki A (2002) A variant of yellow fluorescent protein with fast and efficient maturation for cell-biological applications. Nat Biotechnol 20:87–90

Nguyen QT, Sanes JR, Lichtman JW (2002) Pre-existing pathways promote precise projection patterns. Nat Neurosci 5:861–867

Pauls S, Geldmacher-Voss B, Campos-Ortega JA (2001) A zebrafish histone variant H2A.F/Z and a transgenic H2A.F/Z:eGFP fusion protein for in vivo studies of embryonic development. Dev Genes Evol 211:603–610

Sun Y, Castner EW Jr, Lawson CL, Falkowski PG (2004) Biophysical characterization of natural and mutant fluorescent proteins cloned from zooxanthellate corals. FEBS Lett 570:175–183

Tu H, Xiong Q, Zhen S, Zhong X, Peng L, Chen H, Jiang X, Liu W, Yang W, Wei, J et al (2003) A naturally enhanced green fluorescent protein from magnificent sea anemone (heteractis magnifica) and its functional analysis. Biochem Biophys Res Commun 301:879–885

Valenzuela DM, Murphy AJ, Frendewey D, Gale NW, Economides AN, Auerbach W, Poueymirou WT, Adams NC, Rojas J, Yasenchak J, Chernomorsky R, Boucher M, Elsasser AL, Esau L, Zheng J, Griffiths JA, Wang X, Su H, Xue Y, Dominguez MG, Noguera I, Torres R, Macdonald LE, Stewart AF, DeChiara TM, Yancopoulos GD (2003) High-throughput engineering of the mouse genome coupled with high-resolution expression analysis. Nat Biotechnol 21:652–659

Vintersten K, Monetti C, Gertsenstein M, Zhang P, Laszlo L, Biechele S, Nagy A (2004) Mouse in red: red fluorescent protein expression in mouse es cells, embryos, and adult animals. Genesis 40:241–246

Watanabe T, Costantini F (2004) Real-time analysis of ureteric bud branching morphogenesis in vitro. Dev Biol 271:98–108

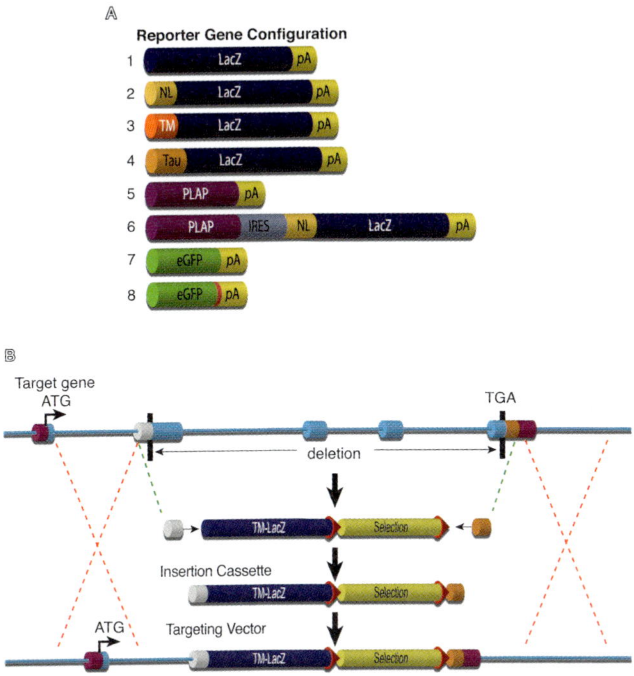

Fig. 7.1. A "Off-the-shelf" reporter gene cassettes. Any of the cassettes *1–8* can be easily swapped in or out of a modularly designed targeting vector to rapidly generate the desired modification in mice. The dual reporter represented by construct *6* attempts to harness all the advantages of β-galactosidase (β-Gal/LacZ) and placental alkaline phosphatase (PLAP) reporters through the use of an IRES sequence. However, despite the appeal of such an approach, it has not been one that we have adopted due to technical limitations that limit its actual utility. *NL* Nuclear localization signal, *TM* transmembrane domain, Tau tubulin binding domain of Tau, *pA* polyadenylation signal, *IRES* internal ribosome entry site; the red band in *8* indicates a membrane-targeting farnesylation signal. **B** Streamlined generation of targeting vectors that employ reporter gene insertion using the VelociGene approach. Short synthetic regions of homology, termed "homology boxes" (small *grey* and *orange cylinders*), corresponding to sequences immediately flanking the region to be deleted in the target gene of interest, are simply ligated to a pre-assembled modular cassette containing the reporter gene of choice (TM-LacZ in this example; *blue cylinder*) and a loxP (*red cones*)-flanked selectable marker (*yellow cylinder*). The homology-box-flanked reporter/selection "insertion cassette" is then reciprocally exchanged for the regions between the homology boxes in a bacterial artificial chromosome (BAC) containing the target gene using homologous recombination in bacteria. The result is essentially a BAC-based targeting vector that can subsequently be used in embryonic stem (ES) cells to make the corresponding modification in the mouse genome. Using this methodology, one is able to readily alter the site of insertion as well as the reporter cassette, to fit the particular goals of a project simply by modifying the short synthetic homology boxes [see (Valenzuela et al. 2003) for a detailed description of this process]

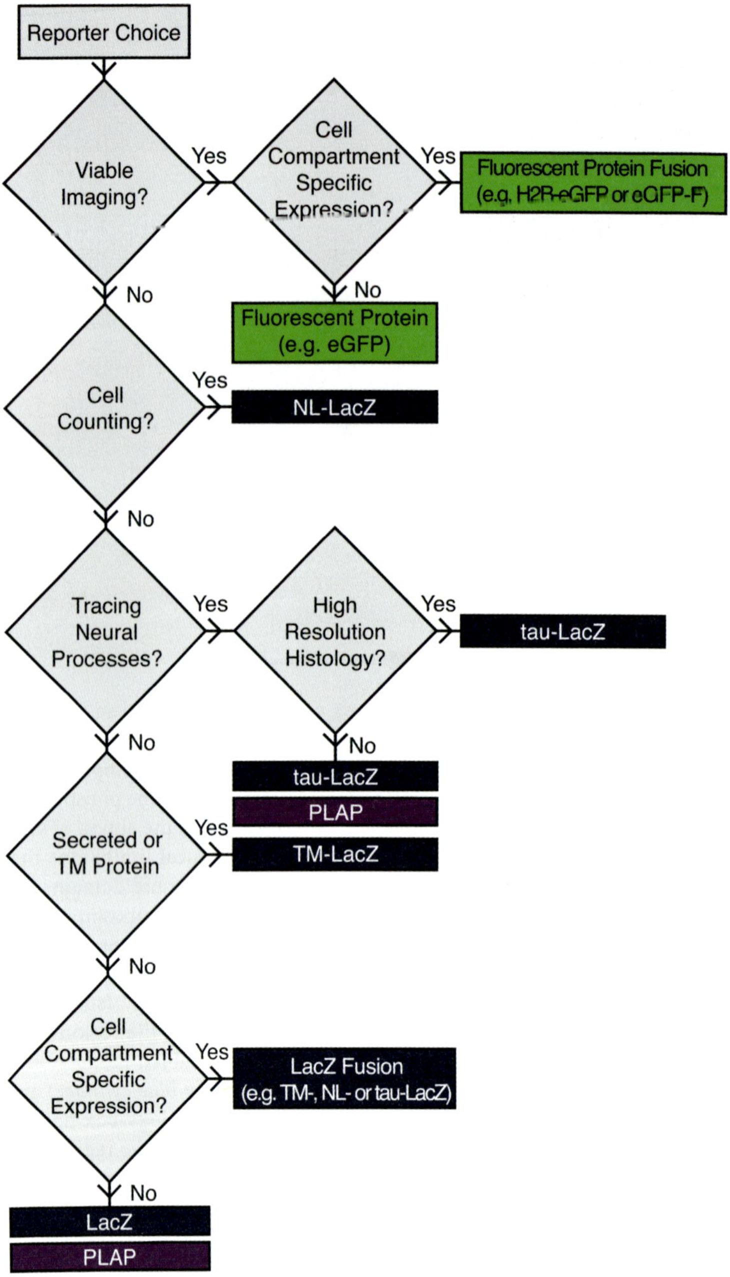

Fig. 7.2. Flow chart to aid in the selection of a particular reporter gene. The questions posed are designed to aid in the decision of which reporter gene to employ in a particular project

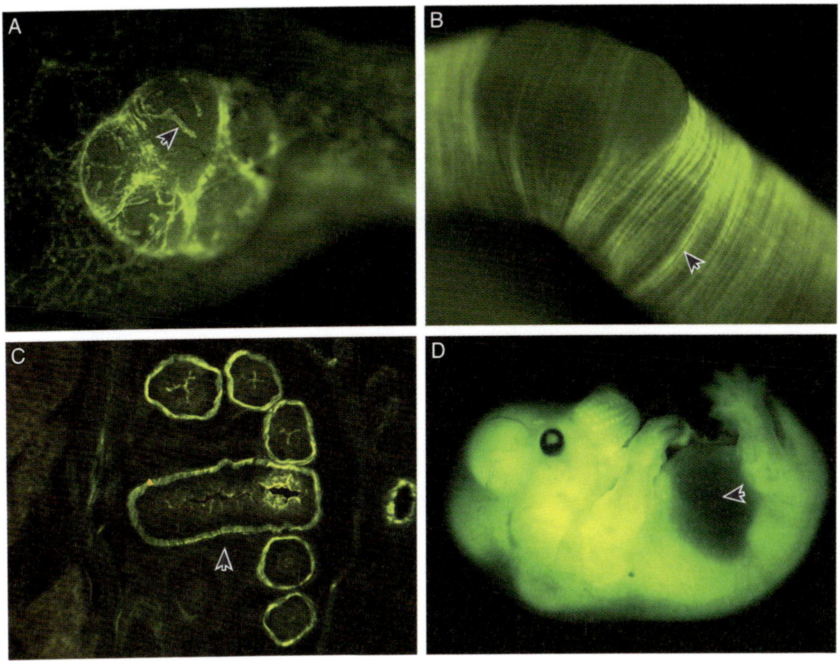

Fig. 7.3. A–D Enhanced green fluorescent protein (eGFP) as a useful marker for revealing gene expression in living tissues. Examples of gene targeting projects in which eGFP has been genetically inserted are shown. **A** eGFP is used to reveal the lymphatic vasculature (*arrowhead*) surrounding a Peyer's patch in the intestine of a living mouse. **B** eGFP expressed in intestinal smooth muscle cells (*arrowhead*). **C** Section through the abdomen of an E15.5 embryo, where eGFP is expressed in the intestinal smooth muscle (arrowhead). **D** eGFP expressed ubiquitously throughout a living E15.5 mouse. Note that liver quenches the fluorescence (*arrowhead*) – the presence of eGFP in the liver was confirmed by Northern analysis in this animal

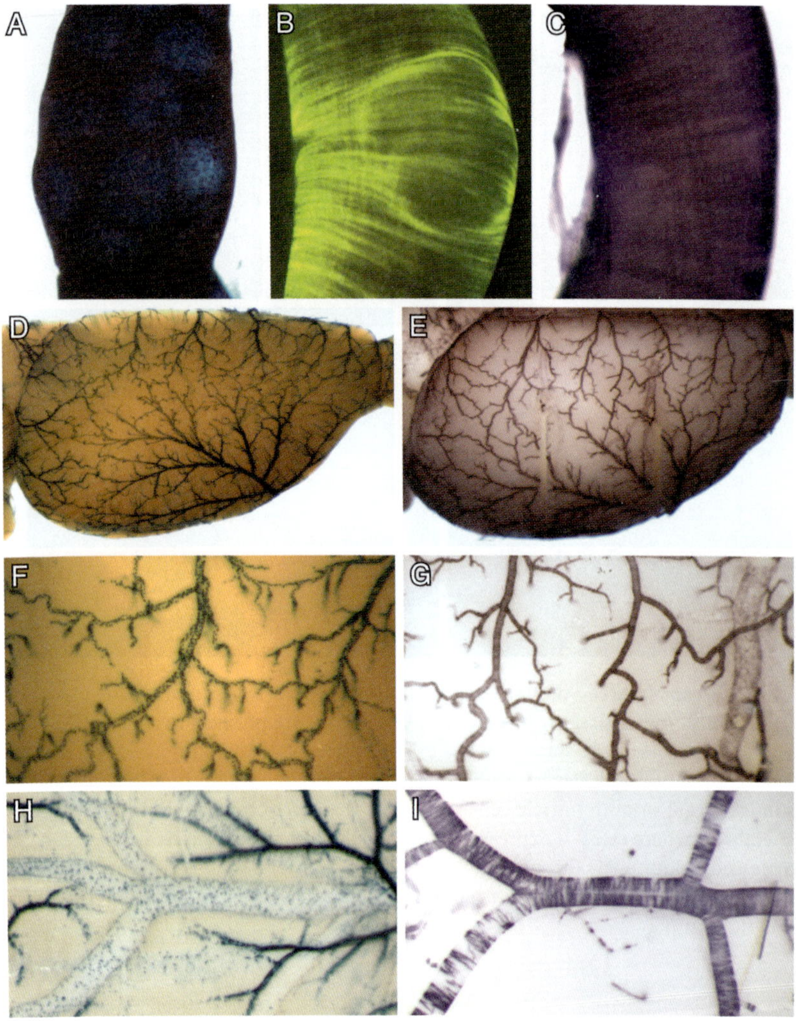

Fig. 7.4. **A–I** Comparison of features of three commonly employed reporter genes, NucLacZ, eGFP and PLAP, each targeted in an identical manner to the same genetic locus, which is expressed in visceral and vascular smooth muscle. **A–C** Detection of NucLacZ (**A**), eGFP (**B**) and PLAP (**C**) expression in the visceral smooth muscle of the intestine. **D** and **F** Low and high power views of NucLacZ expression in the smooth muscle layer of blood vessels in the brain, respectively. Cell nuclei are readily seen and counted in this type of preparation (see **F**). Panels **E** and **G** show the different pattern produced by PLAP expression in equivalent views to **D** and **F**. In the high power view one can readily see the morphology of the smooth muscle cells, which circumferentially wrap the blood vessels. **H** Expression in the nuclei of vascular smooth muscle cells in the diaphragm – individual cells are clearly resolved in veins. **I** PLAP expression in the ventral surface of the brain of an F0 chimera

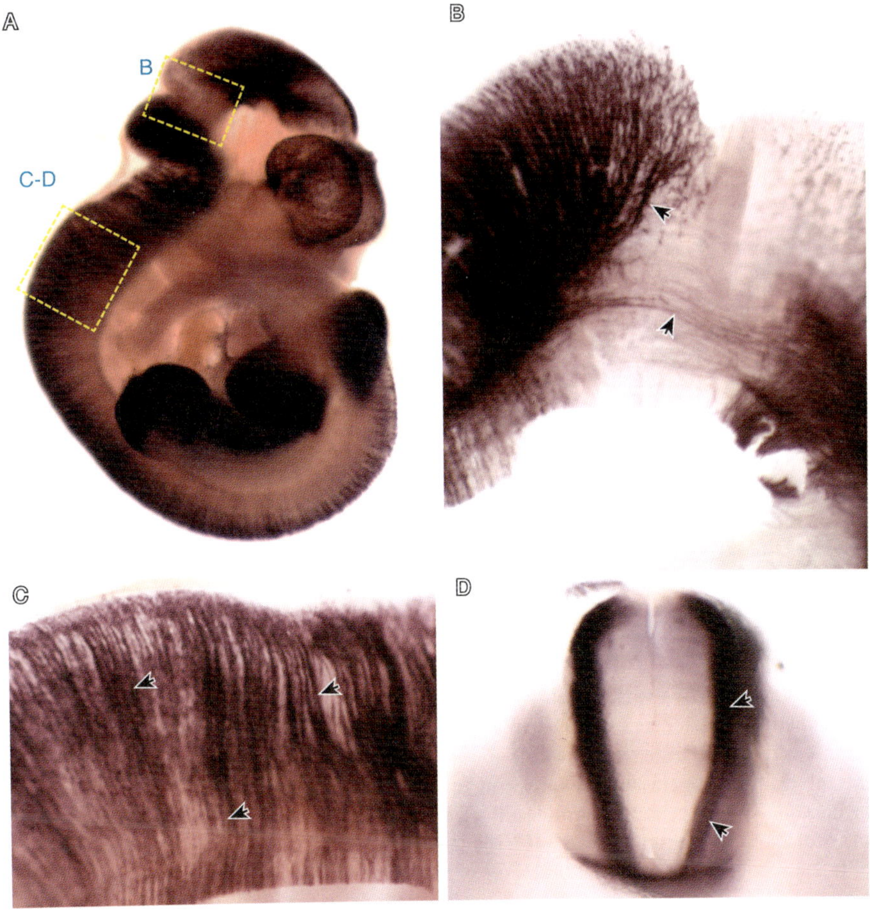

Fig. 7.5. A–D Use of PLAP to reveal cell shapes and processes. When expression is in neurons such as in this example, PLAP staining allows the visualization of elaborate cellular processes, and can even allow the mapping of complex neuronal connectivity. **A** Low power lateral view of a stained E10.5 embryo. The *yellow boxes* roughly indicate from where the images B–D are derived. **B** Lateral view of the dissected hindbrain. **C** Lateral view of the neural tube; note the stained processes of circumferentially projecting commissural axons as revealed by PLAP staining. **D** Cross-section of the neural tube area shown in C, revealing the dorso ventral path of the commissural axons (*arrowheads*)

Fig. 7.6. Examples of a variety of informative patterns of gene expression that can be readily defined using LacZ as a reporter in adult mice. Images **A–I** show the cartilage and bone junction in the rib cage from a whole-mount preparation. Whole tissues are viewed from the interior surface revealing expression in muscle, bone, cartilage, fat, vascular system, or nervous system in these examples. Each image represents an independent gene targeting project. Staining can be seen in cartilage (*arrows*) and intercostal nerves (*arrowheads*); note the resolution of terminal branches of the nerves at the neuromuscular junction *center arrow*) (**A**, **B**); skeletal muscle (*arrowhead*) (**C**); white adipose tissue (*arrowhead*) (**D**, **E**); vascular smooth muscle of arterioles (*arrowhead*); growth plate cartilage is also stained (*arrow*).(**F**). **G–I** Expression in both cartilage (**G**, **H**) and bone (**H**, **J**) components of the rib cage. **H** also has regularly spaced cells in the intercostal regions (*arrowheads*). These cells appear to coincide with major branch points of the intercostal nerves

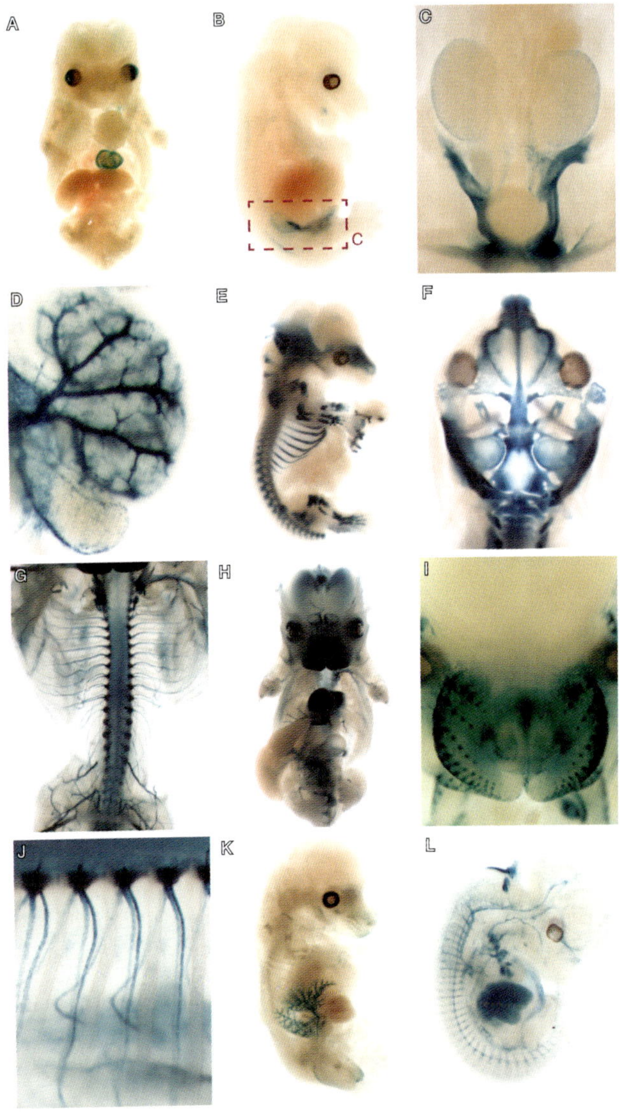

Fig. 7.8. **A–L** Examples of whole-mount gene expression data from E15.5 embryos. Although whole-mount analysis is not generally performed on such late stage embryos, by following the simple dissection shown in Fig. 7.7, one can obtain highly informative and high resolution data relating to domains of expression. **A** Ventro-frontal view of an embryo where β-Gal expression is restricted to the heart. **B** An embryo with two discrete regions of expression: in the developing jaw and gubernaculum (the structure that contracts to cause testicular descent). **C** A closer, ventral view of the region boxed in **B** delineating the expression in gubernaculum. D Staining of the vascular system in the developing kidney. **E, F** Expression in the developing skeleton – the whole embryo (**E**) and dorsal view of the skull (**F**). **G–J** Embryo with β-Gal expression in the developing vasculature and nervous system. **H** Ventro-frontal view. **G** The spinal column of the embryo in **H** with the viscera removed (note the strong expression in the developing nervous system). **J** Close up view of **G** (also derived from the embryo in **H**). **I** Expression in the follicles of the whisker pads on the snout. **K** Strong expression in the developing lungs, seen here in a lateral view of the embryo. **L** Expression in the developing vascular system of a younger embryo

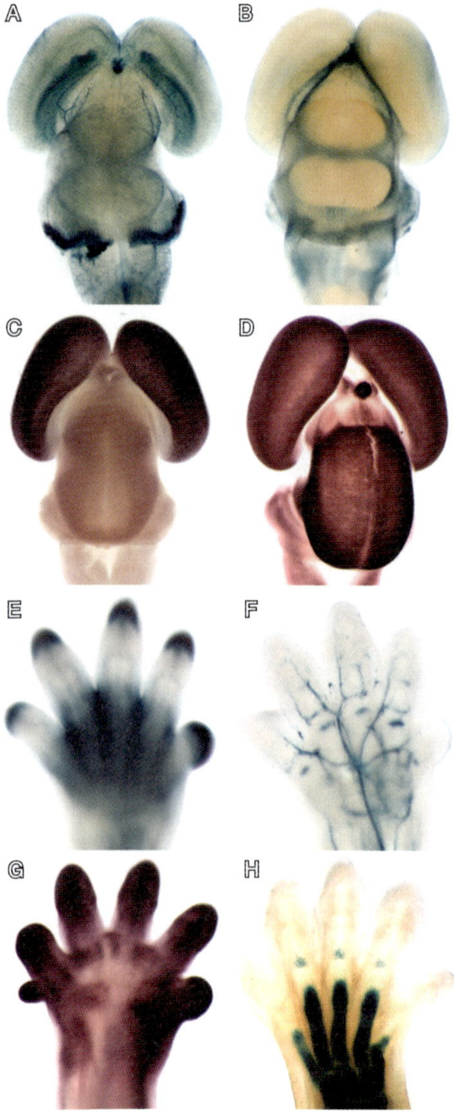

Fig. 7.9. A–H More examples of whole-mount gene expression data obtained from E15.5 embryos. Although the embryos are stained as whole-mounts, good tissue penetration can be achieved when following the dissection guide described here. **A–D** Dorsal views of brains from independent projects that have been sub-dissected following the staining process. **A** and **B** were stained for β-Gal, **C** and **D** for the presence of PLAP. **E–H** Views of the forepaws of embryos (also each from an independent project). **E**, **F** and **H** were stained for the presence of β-Gal, and **G** was stained for PLAP

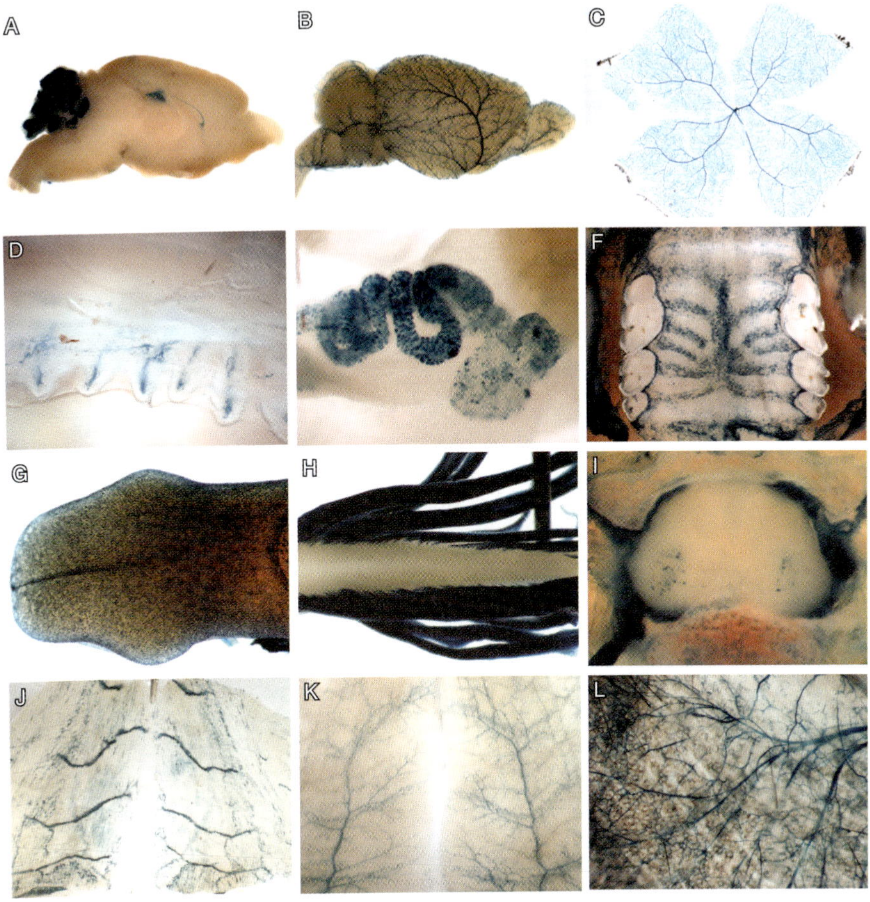

Fig. 7.10. **A–L** Examples of what can be achieved by adult mouse whole-mount staining. Although dissection and processing is designed to screen expression in several animals as rapidly as possible, the process can yield high resolution, publication-quality data. **A** Mid-sagittal view of brain showing expression in cerebellar granule cells, hippocampus and induseum grisium. **B** Lateral view of brain showing strong expression in arterial smooth muscle. **C** Flat-mounted retina showing expression in vasculature. **D** Teeth in the upper jaw showing expression in nerves. **E** Female reproductive organs showing expression in oviducts. **F** Hard palate of the mouth showing expression in oral epithelium. **G** Tongue, dorsal view, showing expression in oral epithelium. **H** Dorsal view of the caudal spinal cord and nerve roots showing expression in Schwann cells. **I** Cut view of the spinal cord and surrounding tissues showing expression in Schwann cells and spinal motor neurons. **J** Ventral abdominal wall showing expression in muscle fascia. **K** Abdominal wall showing expression in vasculature. **L** Ear skin showing expression in peripheral nerves

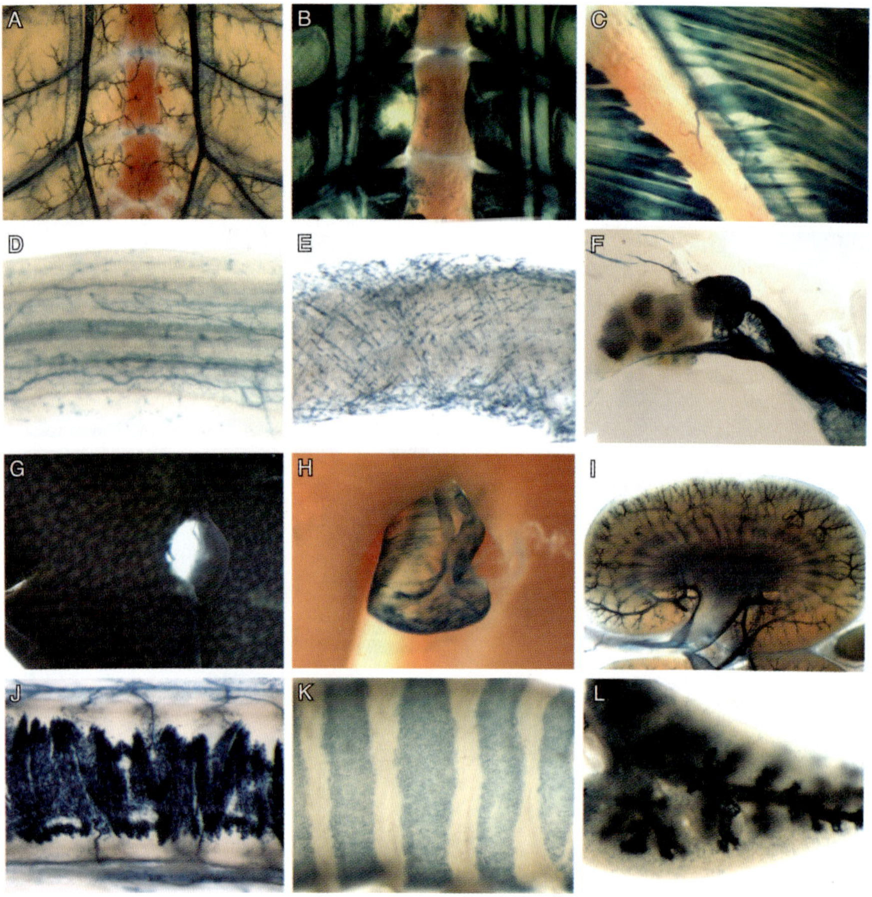

Fig. 7.11. **A–L** More examples of adult mouse whole-mount staining. **A** Internal view of the ventral thoracic wall showing expression in vascular smooth muscle. **B** Internal view of the ventral thoracic wall showing strong expression in white adipose tissue. **C** Intersection of the bone and cartilage elements of the rib showing expression in skeletal muscle and vascular smooth muscle. **D** Esophagus showing expression in vascular endothelium. **E** Esophagus showing expression in skeletal muscle surrounding the esophagus. **F** Female reproductive organs showing expression in ovarian follicles, oviducts, broad ligament, and vasculature in the uterus. **G** Liver and gallbladder showing expression in liver. **H** Liver and gallbladder showing expression in smooth muscle in the gallbladder. **I** Cut view of the kidney showing expression in vasculature, collecting ducts and renal pelvis. **J** Trachea showing expression in vascular and visceral smooth muscle. **K** Trachea showing expression in cartilage. **L** Cut view of lung showing expression in bronchiole epithelium

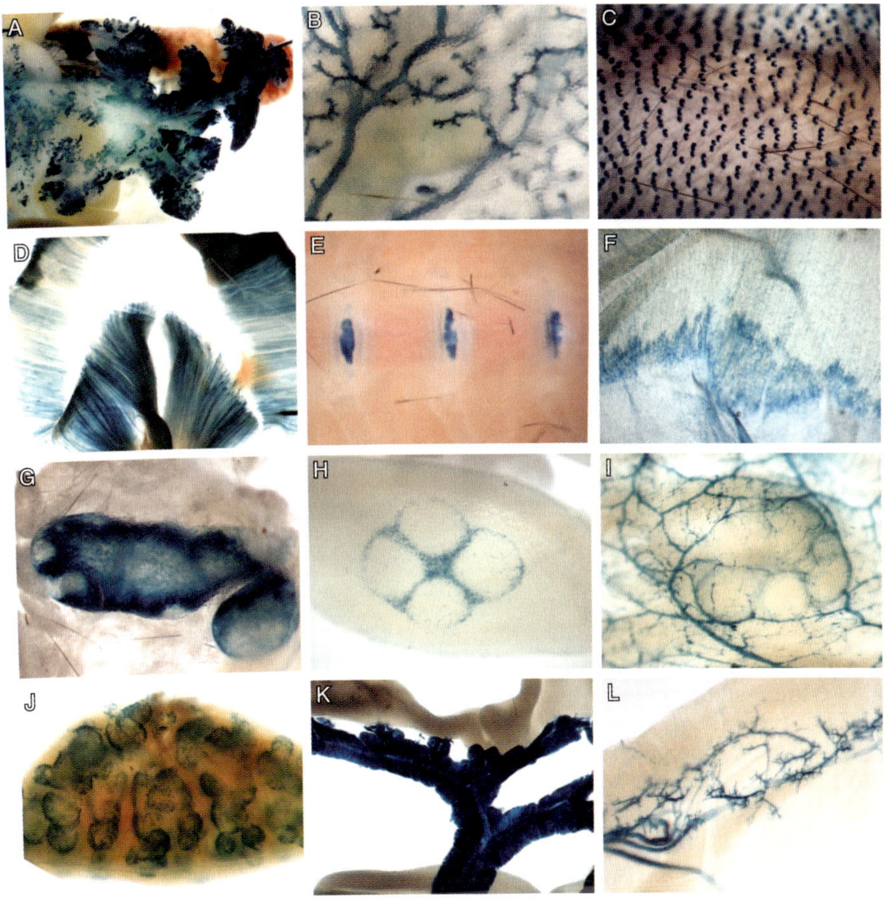

Fig. 7.12. **A–L** More examples of adult mouse whole-mount staining. **A** Pancreas showing expression in exocrine pancreas. **B** Mammary gland with lymph node showing strong expression in epithelium of the mammary ducts. **C** Tail skin showing expression in skin follicles. **D** Diaphragm showing expression in skeletal muscle. **E** Internal view of the dorsal thoracic wall showing expression in vertebral discs of the spine. **F** Diaphragm showing expression in muscle fascia. **G** Mesentery showing expression in lymph node. **H** Small intestine showing expression in the germinal centers of a Peyer's patch. **I** Small intestine showing expression in vasculature of surrounding a Peyer's patch. **J** Cut view of spleen showing expression in white pulp. **K** Mesentery and small intestine showing expression in visceral fat of the mesentery. **L** Diaphragm showing expression in phrenic nerve

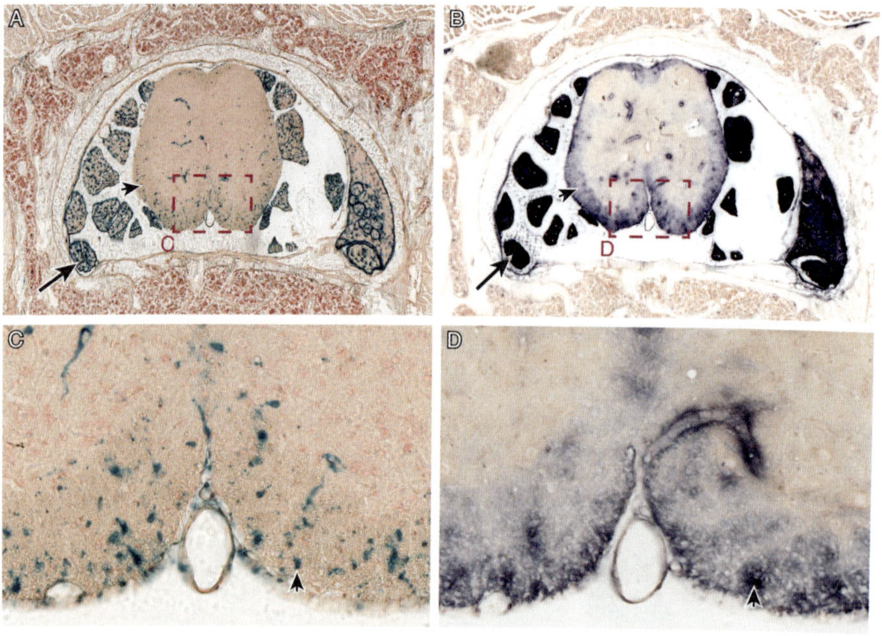

Fig. 7.13. A–D Comparison of standard LacZ with enhanced LacZ staining. Sections shown in **A–D** are from the posterior level of the spine revealing expression in spinal cord and spinal nerves (see Fig. 7.10H for a whole-mount view of a similar region of the spinal cord). The section in **A** (low power view) and **C** (high power view) was stained for 18 h at 37 °C in standard LacZ staining solution. Note that there is only barely detectable staining in the white matter (*arrowheads*), and moderate levels of expression in the spinal nerves (*arrows*) visible. An adjacent section shown in **B** (low power view) and **D** (high power view) was stained for only 2 h at room temperature in the enhanced LacZ staining solution with NBT. The white matter (*arrowheads*) β-Gal activity is now readily detectable. Similarly, expression in the spinal nerves is now very robustly detected (*arrows*). Note, however, that although overall signal intensity is enhanced, the resolution of the signal is diminished by apparent diffusion of the precipitate away from its source

8 Nuclear Transfer in the Cow

William A. Ritchie

8.1 Introduction

Nuclear transfer (NT) is one of the most interesting and challenging areas of science to emerge in recent years. It is a technique that goes back to early experiments in the frog, *Rana pipiens* (Briggs and King 1952), and in the toad, *Xenopus leavis* (Gurdon 1974). The principle of the enucleated egg reprogramming the donor cell to produce identical offspring was continued in mammals (McGrath and Soulter 1983; Willadsen 1986). This technique relied on the use of early embryos as the donor of the nucleus. As the early embryos used were pre-compacted, with 16–32 cells, the number of cells available for the production of cloned animals was limited, although many animals were produced by this method. The next step was to produce animals from the inner cell mass (ICM) of the blastocyst (Smith and Wilmut 1989), but there was still a shortage of cells with which to make clones. One of the greatest breakthroughs was production of a cell line from the ICM. This made possible the production of millions of identical cells for the procedure (Campbell et al. 1996), allowing the production of many identical clones, the cryopreservation of cells for storage, and the potential for the cells to be genetically modified. The story carries on through the birth of Dolly, the first animal cloned from an adult cell, in 1996 (Wilmut et al. 1997) to the present day, with many animals having now been born all over the world by this technique. Animals successfully cloned so far include sheep, cattle, goats, mice, pigs, rabbits and, most recently, cats, although considerable efforts are being made towards the cloning of other species. One recent advance is in the cloning of rare and endangered species such as the Gaur (Vogel 2001), Mouflon (Loi et al. 2001; Ptak et al. 2002) and Enderby Island cattle breed (Wells et al. 1998, 1999). The success rate of NT consistently falls within the range of 1–5%, showing that, at the moment, the technique has an inherent inefficiency. There are many theories as to the reason for this inefficiency but, at present, these remain speculative. Embryos transferred to surrogate mothers are lost in an unusual pattern

William A. Ritchie: Department of Gene Expression and Development, Roslin BioCentre, Midlothian EH25 9PS, Scotland, UK, e-mail: william.ritchie@bbsrc.ac.uk

Principles and Practice
Mammalian and Avian Transgenesis – New Approaches
Shirley Pease and Carlos Lois (Eds.)

in NT, as losses can occur throughout pregnancy and continue after parturition. There have also been defects in many of the animals produced from NT, with reports of animals with defective immune systems, cardiovascular problems, obesity, urogenital abnormalities, and many more problems, many of which are fatal. Dolly herself suffered from arthritis but no one knows whether this was due to the NT process or a chance event in her life.

However, the possible benefits of NT are very great. In particular, the expansion in the number of the available techniques with which to make precise genetic changes in animals by gene targeting in the donor cells, creates many new opportunities in biology, medicine and agriculture. This approach was used to insert genes that direct production of proteins in the milk of livestock, and has since been used to delete a gene from pigs with the aim of being able to use the organs for transplantation into human patients. Genetic selection of the nuclear donor leads to production of genetically modified clones of the desired gender and appropriate production traits. The purpose of this chapter is to give details of procedures that have been used in our laboratory.

8.2 Tools and Equipment

8.2.1 Micromanipulation

An inverted Nikon Eclipse TE300 with differential interference contrast (DIC; Nomarski) optics and epi-fluorescence is used (see Fig. 8.1). This microscope is equipped with 4X, 10X, 20X phase contrast, and 40X DIC lenses and 10X eyepieces. The microscope is fitted with a long working distance condenser, which allows extra room to work on the microscope stage. It is fitted with two Narishige MO-188NE three-dimensional hydraulic oil micromanipulators, which are of the hanging joystick type (www.narashige.co.jp). Two Narishige microinjectors of model IM-16 or IM-188 are used, these being fitted with either a 250 μl gas-tight Hamilton syringe for the holding side of the micro tools or a 100 μl gas-tight Hamilton syringe for the enucleation side. These microinjectors are fitted with a three-way stop-cock (Vigon VG1) and a 5 ml syringe to allow the hydraulic system to be filled easily without having to dismantle the system. The system is filled with Fluorinert (FC77), which is an inert hydraulic fluid, take care to eliminate any air bubbles from the system, as these act like a spring making the injectors difficult to control. The microinjectors are connected to tool holders by polytetrafluoroethylene (PTFE) tubing, the left injector controlling the right tool and the right injector controlling the left tool.

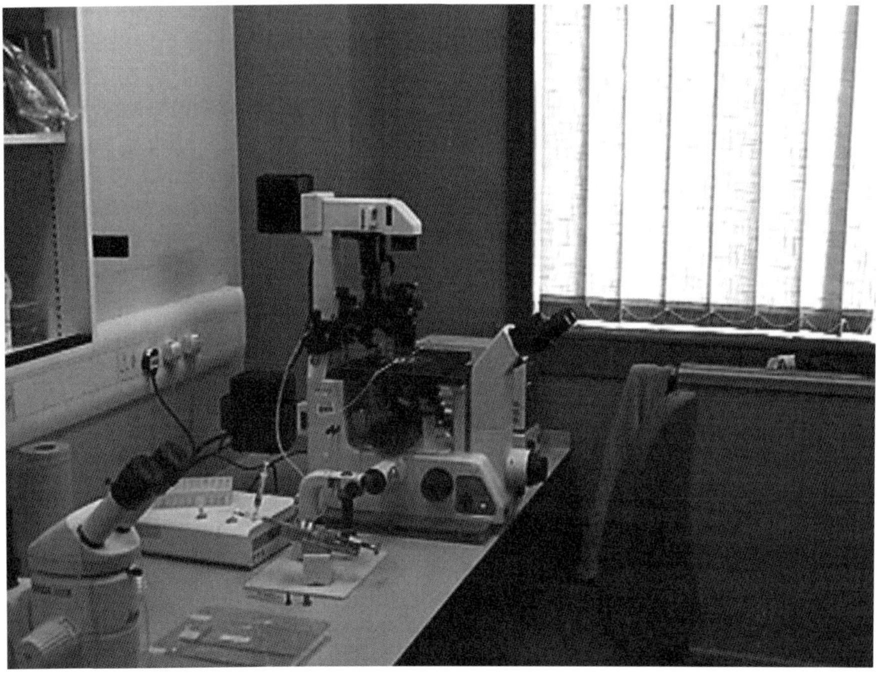

Fig. 8.1. Nikon inverted microscope fitted with Narishige micromanipulators and microinjectors

This system allows both movement of the joystick and manipulation of the suction on the micro tool to be controlled simultaneously.

8.2.2 Microtools

Glass microtools for the oocyte-holding pipette are either purchased from Humagen (Charlottesville, VA) or can be custom made as follows. The holding pipette is made by pulling a 1 mm glass capillary (Harvard Apparatus, cat. no. GC10 100; www.harvardapparatus.com) over a small Bunsen flame to give an outside diameter of 150 μm. The pipette is then bent and the end fire-polished to give an internal diameter of 20 μm, which will allow it to enter horizontally into the manipulation chamber. The enucleation pipette is straight with a long taper and a bevel of 35° and spike, with an external diameter of around 18 μm.

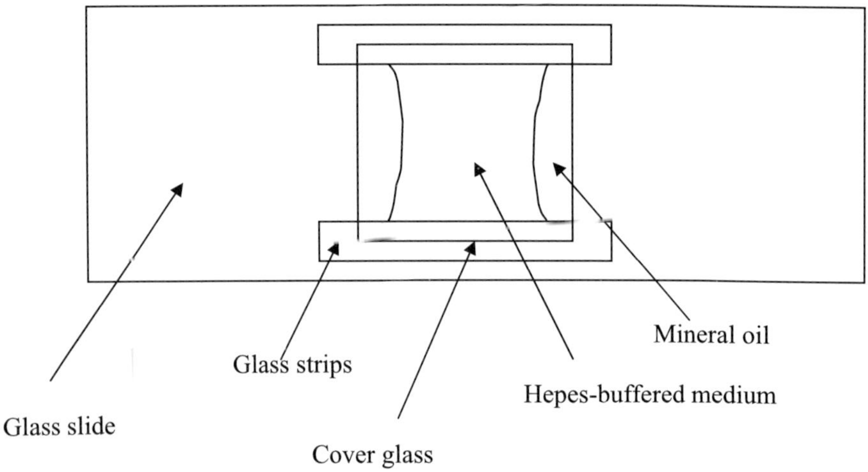

Fig. 8.2. Manipulation chamber

8.2.3
Manipulation Chamber

The manipulation chamber (Fig. 8.2) is made by taking a sterile, good quality, siliconized glass slide and placing two strips of sterile glass, $2 \times 5 \times 25$ mm, on the slide, sealing the strip down with petroleum jelly (Vaseline Ponds) to make it liquid tight, then applying a little petroleum jelly on the upper surface of the glass strips. A 300 μl drop of manipulation medium (hSOF – Ca +FCS, see below) is then placed between the glass strips and a clean sterile glass coverslip is placed on the top, making a liquid-tight seal with the petroleum jelly. The ends of the chamber are then sealed with mineral oil to make a liquid-tight chamber, which will maintain the osmolarity of the solution in the chamber.

The manipulation chamber can then be placed on the microscope stage and the glass microtools inserted from either side. Oocytes or cells can be inserted into the chamber using a small glass capillary, which is pushed through the oil.

8.2.4
Oocyte Handling

Oocytes are handled on Leica Mz 7.5 or Mz 12.5 dissecting microscopes with Linkum MS 100 warm stages or Mini Tüb HT 200 temperature control units (www.minitube.de). The temperature is held at 37 °C.

8.2.5 Cell Fusion

Cells are fused with a BSL CF-150 cell fusion device (Biological Laboratory Equipment, Budapest, Hungary). This is connected to a custom-made fusion chamber consisting of two platinum electrodes, which are stuck to the bottom of a glass Petri dish with epoxy resin adhesive (Araldite), at a distance of 200 μm apart.

8.2.6 Embryo Culture

NT embryos are cultured in a Heraeus BB6060 (Hanau, Germany) triple gas incubator with a gas mixture of 5% O_2, 5% CO_2, 90% N_2, at 39 °C in a humidified atmosphere.

8.2.7 Supplies

All chemicals (and catalogue numbers) are from Sigma (St. Louis, MO), unless stated otherwise.

- Fluorinert FC 77 (F-4758); $NaHCO_3$ (S-5761); Phenol Red (P-5530); pyruvic acid (P-4562); $CaCl_2 \cdot 2H_2O$ (C-7902); glucose (G-6152); $MgCl_2 \cdot 6H_2O$ (M-2393); NaCl (S-5886); KCl (P-5405); KH_2PO_4 (P-5655); kanamycin sulphate (K-1377); Penicillin (P-7794); Streptomycin (S-1277); L-glutamine (G5763); Hepes, free acid (H-6147); Hepes, sodium salt (H-3784); sodium lactate (60%) syrup (L-7900); bovine serum albumin (BSA), fatty-acid-free (FAF) (A8806); non essential amino acids (M-7145); essential amino acids (B-6766); cytochalasin B (CB; C-6762); hyaluronidase (H-3884); bis benzimide (Hoechst 33342; B-2261); mannitol (M-4125), dimethyl sulfoxide (DMSO; D-4540), Sigmacote (SL-2); estradiol (E-2758); leutinizing hormone (LH; L-5269); $MgSO_4 \cdot 7H_2O$ (M-1880); mineral oil, embryo tested (M-8410)
- TCM199 with Earles Salts (Gibco, Paisley, UK)
- Fetal calf serum (FCS)
- Follicle stimulating hormone (FSH) (Ovagen ICP, Auckland, NZ)
- 18 μm bevelled and sharpened pipettes. Humagen (Charlottesville, VA)

8.3 Media and Solutions

8.3.1 Aspiration Medium

- Hepes-buffered TCM199
- 10% FCS, heat inactivated

8.3.2 Dissection Medium (50 ml)

- 45 ml TCM199 with Earles salts (Gibco)
- 75.0 mg/l kanamycin monosulfate
- 7.08 g/l Hepes (pH 7.8 osmolarity 279 mosmol/kg H_2O)
- 5 ml FCS (heat inactivated)

8.3.3 Maturation Medium

For 10 ml:

- 8.89 ml Bicarbonate-buffered TCM199
- 1 ml FCS
- 0.1 ml LH stock (1 μg/ml final concentration)
- 0.1 IU FSH (Ovagen) 0.9 mg
- Filter sterilize then add
- 0.01 ml estradiol (1 μg/ml final concentration)
- Equilibrate at 39 °C in 5% CO_2 for 2 h
- Add 0.1 ml 10 mM cysteamine (make up weekly and store at 4 °C)

8.3.4 Stock solution B (250 mM $NaHCO_3$)

- 1.05 g $NaHCO_3$

- Phenol Red (2–3 grains)
- Make up to 50 ml with MilliQ dH_2O (Osmolarity 430–440)
- Store filtered at 4 °C for 3 weeks

8.3.5 Stock solution C (33 mM pyruvate)

- 0.036 g pyruvic acid
- Make up to 10 ml with MilliQ dH_2O (Osmolarity 55–60)
- Store filtered at 4 °C for 1 week

8.3.6 Stock solution D (171 mM $CaCl_2 \cdot 2H_2O$)

- 1.26 g $CaCl_2 \cdot 2H_2O$
- Make up to 50 ml with MilliQ dH_2O
- Osmolarity 430–440
- Store filtered at 4 °C for 3 months

8.3.7 Stock solution G (60 mM Glucose)

- 0.54 g glucose
- Make up to 50 ml with MilliQ dH_2O
- Store filtered at 4 °C
- Osmolarity 60–70
- Store filtered at 4 °C for 3 months

8.3.8 Stock solution GLN (10 mM L-Glutamine)

- 0.0438 g L-Glutamine
- Make up to 30 ml with dH_2O
- Store filtered at 4 °C for 1 week

8.3.9
Stock solution H (250 mM Hepes)

- 3.0 g Hepes (free acid)
- 3.25 g Hepes (sodium salt)
- Phenol Red (2–3 grains)
- Make up to 100 ml with MilliQ dH_2O
- Osmolarity 380–385
- Store filtered at 4 °C for 1 month

8.3.10
Stock solution K (kanamycin sulphate)

- 0.5 g kanamycin sulphate
- Make up to 10 ml with MilliQ dH_2O
- Store filtered at 4 °C for 3 months
- Dilute 1:1,000 for a final concentration of 50 μg/ml

8.3.11
Stock solution L (330 mM Na lactate)

- 11.75 ml Na lactate (60%) syrup
- Make up to 250 ml with MilliQ dH_2O
- Osmolarity 590–640
- Store filtered at 4 °C for 3 months

8.3.12
Stock solution M ($MgCl_2 \cdot 6H_2O$)

- 2.0 g $MgCl_2 \cdot 6H_2O$
- Make up to 200 ml with MilliQ dH_2O
- Osmolarity 120–130
- Store filtered at 4 °C for 3 months

8.3.13 Stock solution S2

- 6.29 g NaCl
- 0.534 g KCl
- 0.162 g KH_2PO_4
- 0.025 g Kanamycin (or 0.06 g Penicillin + 0.05 g Streptomycin)
- Make up to 100 ml with MilliQ dH_2O
- Osmolarity 2173
- Store filtered at 4 °C for 3 months

8.3.14 Hepes synthetic oviduct fluid (hSOF) +Ca

	To make 100 ml
Stock 2	10.0 ml
Stock B	0.2 ml
Stock H	8.0 ml
Stock C	1.0 ml
Stock D	1.0 ml
Stock M	1.0 ml
Stock L	1.0 ml
Stock G	2.5 ml
MilliQ H_2O	73.5 ml

- Osmolarity 265–275
- pH 7.4
- Add 0.3 g BSA (FAF)
- Filter and store at 4 °C for 1–2 weeks

8.3.15
Hepes SOF (hSOF) –Ca

	To make 100 ml
Stock 2	10.0 ml
Stock B	2.0 ml
Stock H	8.0 ml
Stock C	1.0 ml
Stock M	1.0 ml
Stock L	1.0 ml
Stock G	2.5 ml
MilliQ H_2O	74.5 ml

- Osmolarity 265–275
- pH 7.4
- Add 0.3 g BSA (FAF)
- Filter and store at 4 °C for 1–2 weeks

8.3.16
Culture Medium (SOFaaBSA)

	To make 100 ml
Stock 2	10.0 ml
Stock B	10.0 ml
Stock C	1.0 ml
Stock D	1.0 ml
Stock M	1.0 ml
Stock L	1.0 ml
Stock G	2.5 ml
Stock GLN	10.0 ml
Non essential amino acids	1.0 ml
Essential amino acids	2.0 ml
MilliQ H_2O	60.5 ml

- Osmolarity 265–275
- pH 7.4
- Add 0.8 g BSA (FAF)
- Filter and store at 4 °C for 1 week

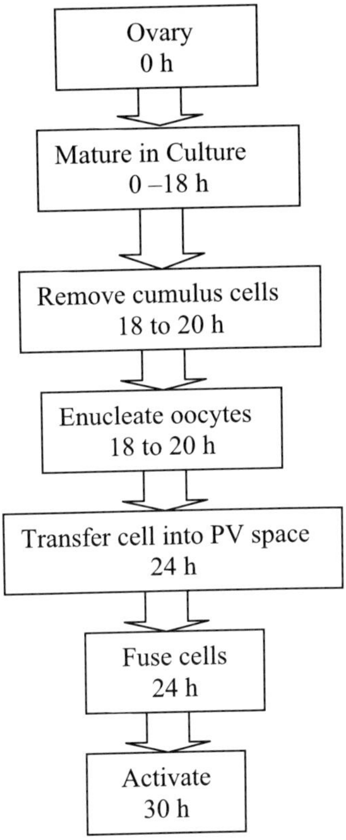

Fig. 8.3. Schematic representation of nuclear transfer (NT) procedure. PV Perivitelline

8.3.17 Fusion Medium

- Mannitol 0.3 M
- $MgSO_4 \cdot 7H_2O$ 0.1 mM
- $CaCl_2 \cdot 2H_2O$ 0.05 mM
- Filter and store at −20 °C
- Note: Use a 0.2 μm syringe filter

8.4 Procedure

A schematic representation of an NT procedure is shown in Fig. 8.3.

8.4.1 Oocyte Maturation

- Reagents required:
  - PBS
  - Dissection medium
  - FCS
  - Maturation medium
  - Oil

1. Oocytes are prepared by collection of ovaries in a warm container following slaughter of heifers.
2. The ovaries are transported to the laboratory, in warm PBS, within 4 h of slaughter and should be between 25 and 35 °C when they reach the laboratory.
3. The ovaries are thoroughly washed in warm saline to remove excess blood before follicles are aspirated with an 18-gauge needle and 10 ml syringe. Note. Only follicles between 4 and 8 mm should be selected for aspiration.
4. The follicles are pierced and the cumulus oocyte complexes (COCs) removed in the follicular fluid by applying gentle suction on the piston of the syringe. Each follicle contains one COC, and the number of follicles per ovary is variable, some producing one or two and others up to four or five.
5. The aspirated follicular fluid containing the COCs is gently expelled into a universal tube and kept warm in a water bath until the ovaries have been aspirated.
6. The COCs are allowed to settle for 10 min before the excess follicular fluid is carefully aspirated from the top of the tube using a sterile plastic pipette.
7. The follicular fluid is diluted with an equal volume of dissection medium +10% FCS, then poured into a large Petri dish. The tube is rinsed with a few milliliters of dissection medium +FCS, which is then added to the Petri dish.
8. A search is made for COCs that have at least three layers of cumulus cells surrounding the oocyte. COCs with fewer cumulus cells are

either degenerating or have had the cells physically removed during collection.

9. Selected COCs are washed twice in dissection medium and then once in maturation medium. A pipettor is used to move the media around, so that the cumulus cells are not scraped off.
10. Culture ten oocytes/50 μl drop maturation medium under oil in a 5% CO_2 in air mixture at 39°.

8.4.2 Enucleation of Oocytes

- Reagents required:
  - hSOF –Ca
  - FCS
  - Hyaluronidase
  - Nunc four-well plates (www.nalgenenunc.com)
  - CB
  - Hoechst 33342 dye
  - BSA (FAF)

1. After 18–20 h of maturation, gently pipette up and down a few times to remove any free cumulus.
2. The COCs are washed in hSOF –Ca without FCS before transfer to a well with 300 IU hyaluronidase in hSOF –Ca –FCS and then gently pipetted using an automatic pipettor until all of the cumulus cells have been removed. A 100 or 200 μl volume is sucked into the pipette from a 500 μl volume in a Nunc four-well plate. Keep the pipette at an angle so as the oocytes are not damaged by being squeezed against the bottom or side of the well.
3. Wash in hSOF –Ca +10% FCS.
4. Oocytes with a first polar body are selected under a dissecting microscope by gently rolling the oocytes until the polar body is seen. These oocytes, which are thought to be mature and at MII, are then stored in hSOF –Ca +10% FCS in a warm box at 37 °C until required.
5. Oocytes put to one side as not having a polar body can be examined again later.
6. Batches of 10–15 oocytes are placed in hSOF –Ca plus 7.5 μg/ml CB and 10 μg/ml Hoechst 33342 for 15 min prior to enucleation.

7. Oocytes are placed in the manipulation chamber with hSOF –Ca +FCS with CB.
8. Move the holding pipette to the oocyte and apply gentle suction until an oocyte is picked up.
9. This oocyte is then raised above the rest of the oocytes, and the microscope stage moved to an area of the chamber away from any other oocytes.
10. Turn the oocyte using the enucleation pipette until the polar body is in a position where the pipette can be pushed through the zona pellucida, the polar body and the adjacent cytoplasm within the body of the oocyte (where the metaphase plate is located), such that the polar body and metaphase plate can be aspirated gently into the pipette (Fig. 8.4). An 18 μm pipette ground at 35° with a pipette beveller and a spike put on the end with a hot glass bead on a microforge to allow it to go easily through the zona pellucida, is used.
11. The holding pipette with oocyte attached is removed from the field of view; the white light is turned off before the enucleation pipette is exposed to UV light. The DNA in the polar body will be seen to fluoresce brightly, with the maternal DNA fluorescing less brightly (Fig. 8.4). Depending on the orientation of the metaphase plate, the chromosomes can be seen as circular, linear or any configuration in between. *Note:* it is important that the enucleated oocyte is not exposed to UV light.
12. Successfully enucleated oocytes are deposited to the right of the chamber so that they can be easily removed, using a small hand-drawn glass pipette.
13. Wash enucleated oocytes in hSOF –Ca +FCS. Store in the incubator in SOF –Ca +BSA (0.8 mg/ml FAF BSA). *Note:* enucleation of the oocyte is most easily carried out as soon as the polar body is expelled, as the metaphase lies near the polar body at this time. Therefore the polar body can be used as a marker for enucleation. Polar bodies are expelled from around 18 h of maturation onwards, so enucleation can begin at around 20 h post maturation. Electrofusion can then be carried out at 24 h.

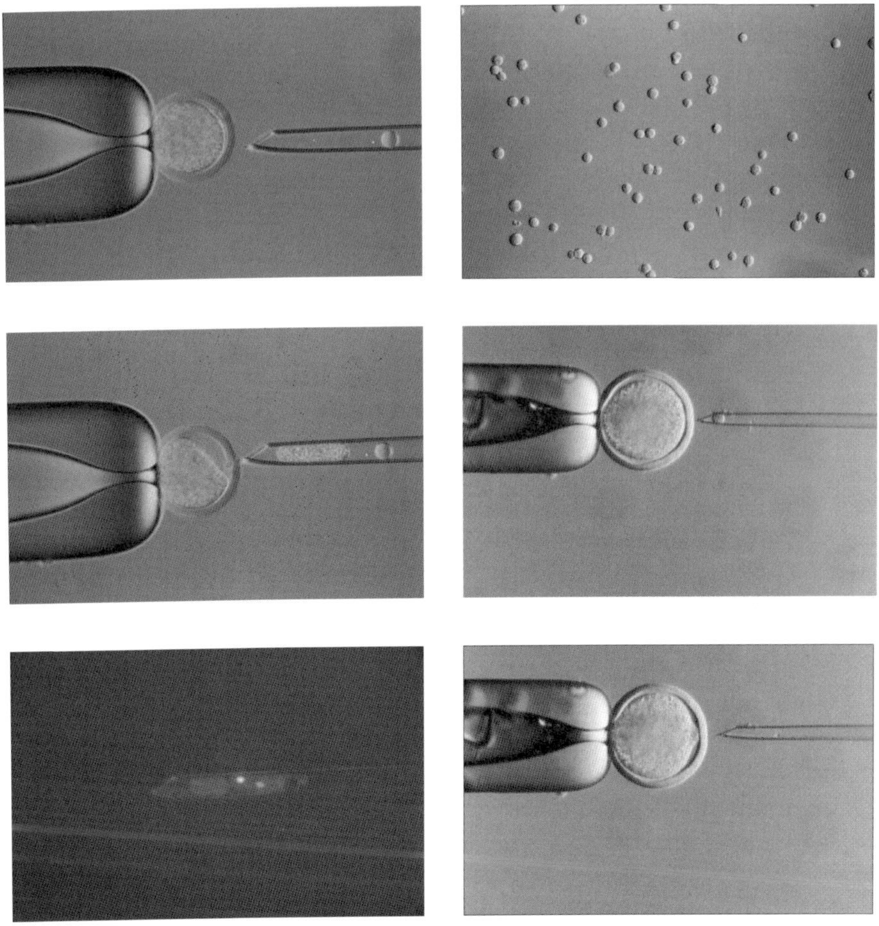

Fig. 8.4. Enucleation and cell transfer. Left: *Top* Oocyte orientated ready for enucleation, *middle* cytoplasm and polar body removed from the oocyte, *bottom* pipette exposed to UV light to show metaphase chromosomes and polar body. Right: *Top* Individual cultured cells ready for transfer, *middle* cell in transfer pipette, *bottom* cell transfer to enucleated oocyte

8.4.3 Cell Preparation

- Reagents required:
  - Donor cell culture medium
  - Trypsin
  - hSOF
  - FCS

1. Donor cells are prepared by serum starving cells in culture before they reach confluence by reducing the serum from 10% to 0.5% for 5 days. This causes the majority of the cells to enter quiescence (G0). (*Note:* cells used for NT have often been serum starved to induce quiescence and this enhances the efficiency of the procedure. Allowing the cells to become confluent may also aid efficiency, since this also results in cell quiescence)
2. The cells are trypsinized until they are in a single cell suspension, and then washed and re-suspended in hSOF +0.5% FCS medium until injection (Fig. 8.4).

8.4.4 Injection of cells

- Reagents required:
  - hSOF –Ca
  - FAF BSA

1. Cells are transferred to the top right-hand corner of a fresh manipulation chamber with hSOF –CA +BSA (0.8 mg/ml FAF BSA).
2. Batches of ten enucleated oocytes are transferred to the center of the chamber using a small hand-drawn pipette.
3. Oocytes are picked up using the holding pipette.
4. A number of cells are then picked up in the injection pipette and a single intact cell is injected through the zona pellucida and into the perivitelline space of each of the enucleated oocytes. Care must be taken to ensure that the cell is alive and is making good contact with the oocyte cytoplasm for fusion to take place. Live cells have a bright shiny cell membrane, but if they lyse during cell transfer, the cell membrane becomes dull.

8.4.5 Electrofusion

- Reagents required:
  - Fusion medium
  - hSOF –Ca
  - FCS
  - FAF BSA
  - oil

1. The couplets are removed from the manipulation chamber, washed in fusion medium (mannitol and calcium), and placed in the fusion chamber.
2. The couplets are all placed to one side of the electrodes, and each couplet is lifted in turn, using a small glass pipette, and placed between the electrodes.
3. Couplets are manually aligned so that the point of contact between the cell and oocyte is parallel to the electrodes. Precise orientation of the couplet is required for fusion to take place (Fig. 8.5).
4. An alternating current (AC) current of 0.25 kV/cm is applied for a period of 5 s, and followed by 3× direct current (DC) pulses of 1.25 kV/cm of 80 μs in rapid succession.
5. Couplets are removed from the fusion chamber and deposited in a wash dish containing hSOF –Ca +FCS.
6. Upon completion of this procedure, the batch of couplets are removed from the wash and transferred to 30 μl drops of SOF –Ca +BSA under oil and incubated at 39 °C in 5% CO_2 in air.
7. Check if fusion has taken place after 1 h.
8. Up to 100 oocytes can be handled comfortably each manipulation day.

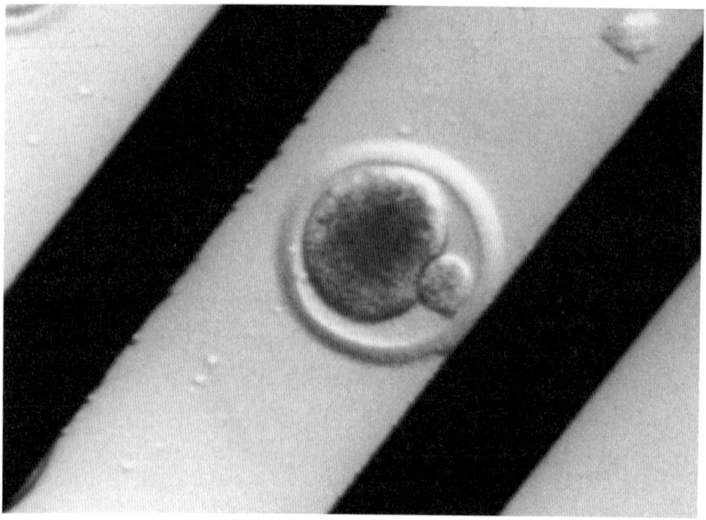

Fig. 8.5. Enucleated oocyte with cell in the PV space, between two electrodes

8.4.6 Activation

- Reagents required:
  - hSOF+Ca
  - FAF BSA
  - Ionomycin
  - SOFaaBSA
  - 6-dimethylaminopurine

1. 30 min before activation wash in hSOF +Ca + BSA(1 mg/ml FAF BSA)
2. At 26–30 h post-maturation, the embryos are activated by placing them in hSOF +Ca +BSA containing 5 μM ionomycin at room temperature for 5 min.
3. The embryos are repeatedly washed in hSOF +Ca +BSA (30 mg/ml FAF BSA) to remove the ionomycin.
4. Culture for 4 h in SOFaaBSA (0.8 mg/ml FAF BSA) with 2 mM 6-dimethylaminopurine (6-DMAP) at 39 °C in a 5% O_2:5% CO_2:90% N_2 gas mixture in drops under oil.
5. Wash in pre-gassed SOFaaBSA (0.8 mg/ml FAF BSA)
6. Culture until the blastocyst stage is reached.

8.4.7 Culture

- Reagents required:
  - SOFaa
  - FAF BSA

1. The successfully fused and activated embryos are transferred to 500 μl SOFaa+BSA, overlaid with mineral oil, in Nunc four-well plates at 39 °C in a 5% O_2:5% CO_2:90% N_2 gas mixture.
2. Embryos should have developed to become blastocysts on day 7 of culture. *Note:* The success of NT is dependent on a large numbers of variables, from the treatment of the heifers prior to slaughter, to the "cell line", which is used as the nuclear donor, so it is very difficult to give accurate figures for the success of the procedure.

- Cloning statistics:
  - 14–69% reach morula/blastocyst stage, as a percentage of embryos enucleated and reconstructed.
  - 0–100% pregnant, as a percentage of the number of pregnancies detected to the number of recipient animals.
  - 0–83% live births, as a percentage of embryos transferred.
  - 0–2.8% cloning efficiency, as a percentage of live offspring from those enucleated and reconstructed.

8.5 Troubleshooting

As mentioned earlier, NT is a procedure involving a great many steps, many of which are critical to the success of the technique. Oocyte quality is very important, and it is wise to incorporate some controls into the procedure to make sure that the oocytes are of good quality and that they are at the correct stage of development. This is the rationale behind selecting oocytes with polar bodies, as these are liable to be at the MII stage.

It may also be wise to make sure that the oocytes are at the correct stage for activation by activating a sample at different times following maturation.

Routine maintenance of equipment and recording of temperatures and gas concentrations of incubators can prevent problems.

Make sure that the holding pipette is lifted above the level of the other oocytes when moving the pipette around as it can easily move the other oocytes in the chamber if it is not lifted high enough.

The area where enucleation is carried out should be a sufficient distance from the other oocytes and the enucleated oocyte on the holding pipette, to ensure that only the enucleation pipette is exposed to UV radiation.

When transferring cells to the perivitelline space of the enucleated oocyte, care must be taken to ensure that the cytoplasm and cell are in contact, so that electrofusion can take place.

If using cultured cells, karyotype the cell line at regular intervals to make sure that the cells have not developed an abnormal karyotype during culture.

Different cells require different periods of serum starvation to trigger quiescence, and this can be determined by proliferating cell nuclear antigen (PCNA) staining. PCNA is a nuclear antigen visible during S-phase of the cell cycle.

8.6 Discussion

NT is a very labor intensive and skilled procedure that has always required a great deal of manual dexterity, but there are methods now in the pipeline that may make the procedure easier. Bisecting oocytes and fusing two halves, without chromosomes, together, along with a somatic cell has already shown promise, requires less sophisticated machinery and much less skill. There is also the possibility of mechanization and increased efficiency in the future.

Acknowledgements. I would like to thank Prof. Ian Wilmut in whose laboratory the experiments on Nuclear Transfer have been carried out, and also all of my friends and colleagues who have assisted me with these experiments. I acknowledge the support of the sponsors listed in all of our papers and I am grateful for the financial support of the Government, industry and commercial organizations.

References

Briggs R, King TJ (1952) Transplantation of living nuclei from blastula cells into enucleated frog's eggs. Proc Natl Acad Sci USA 38:455–461

Campbell KHS, McWhir J, Ritchie WA, Wilmut I (1996) Sheep cloned by nuclear transfer from a cultured cell line. Nature 380:64–66

Gurdon JB (1974) The control of gene expression in animal development. Clarendon Press, Oxford

Loi P, Ptak G, Barboni B, Fulka J Jr, Cappai P, Clinton M (2001) Genetic rescue of an endangered mammal by cross species nuclear transfer using post mortem somatic cells. Nat Biotechnol 19:962–964

McGrath J, Soulter D (1983) Nuclear transplantation in the mouse embryo by microsurgery and cell fusion. Science 220:1300–1302

Ptak G, Clinton M, Barboni B, Muzzeddu M, Cappai P, Tischner M, Loi P (2002) Preservation of the wild european Mouflon: the first example of genetic management using a complete program of reproductive biotechnologies. Biol Reprod 66:796–801

Smith LC, Wilmut I (1989) Influence of nuclear and cytoplasmic activity on the development in vivo of sheep embryos after nuclear transplantation. Biol Reprod 40:1027–1035

Vogel G (2001) Endangered species. Cloned Gaur a short-lived success. Science 291:409

Wells DN, Misica PM, Tervit HR, Vivanco HW (1998) Adult somatic cell nuclear transfer is used to preserve the last surviving cow of the Enderby Island cattle breed. Reprod Fertil Dev 10:369–78

Wells DN, Misica PM, Tervit HR (1999) Production of cloned calves following nuclear transfer with cultured adult mural granulosa cells. Biol Reprod 60:996–1005

Willadsen SM (1986) Nuclear transplantation in sheep embryos. Nature 320:63–65

Wilmut I, Schnieke AE, McWhir J, Kind AJ, Campbell KHS (1997) Viable offspring derived from fetal and adult mammalian cells. Nature 385:810–813

9 Production of Transgenic Pigs by DNA Microinjection

Robert M. Petters, Rebecca L. Krisher

9.1 Introduction

Transgenesis includes a number of steps that are common to all the species discussed in this volume: superovulation, embryo recovery, embryo microinjection (or another method of transferring DNA), and embryo transfer to a recipient animal for gestation to term. Livestock species such as the pig can present special problems because of their size and housing requirements but they may have special advantages, such as litter size (12–14 piglets). The experimental manipulation of pigs for the production of transgenics is very much like that for mice, with some key differences that are primarily logistical in nature. We will try to point these out and to make recommendations where possible.

Transgenic pigs are key players in the area of biomedical models of human disease, and hold the promise of unlimited organs for transplant to humans (Phelps et al. 2003). Refinements based originally on transgenic technology such as cloning and knock-out strategies bring the pig to the forefront of biomedical research (Boquest et al. 2002). Man may live longer with pig parts substituting for failing human organs. The knowledge gained from pigs will bring many advances to human medicine. The techniques described below have been used to make transgenic pigs, and many of them can be used in other more recently developed techniques, such as cloning.

9.1.1 In Vivo Produced Embryos

In vivo produced embryos will always have the greatest viability and thus, almost all experiments have used fertilized embryos from donor females.

Robert M. Petters: Department of Animal Science, North Carolina State University, Box 7621, Raleigh, NC 27695–7621, USA, e-mail: bob_petters@ncsu.edu

Rebecca L. Krisher: Department of Animal Sciences, Lilly Hall of Life Sciences, Lab VPTH 201/204, 915 West State St., Purdue University, West Lafayette, IN 47907–2054, USA

Principles and Practice
Mammalian and Avian Transgenesis – New Approaches
Shirley Pease and Carlos Lois (Eds.)

In our laboratory, we have synchronized and superovulated prepubertal female pigs (Sommer et al. 2002), flushed the embryos from the reproductive tract, and proceeded to manipulate the embryos. Surviving embryos were then transferred to one or two of the best donors for gestation to term. Synchronized, superovulated flushed donors serve as recipients, with very good success. They have been screened for their response (number and quality of embryos) and their ovaries have been evaluated prior to transfer. In terms of viability, in vivo produced embryos should always have the highest viability compared to in vitro produced embryos. Improvements in in vitro procedures will narrow the gap between in vitro and in vivo embryos, but will probably never close it.

9.1.2 In Vitro Produced Embryos

Matured porcine oocytes can be purchased from a number of commercial sources. These oocytes can be brought into the laboratory and subjected to in vitro fertilization to provide embryos for microinjection (Abeydeera 2001). However, there are some problems. First, in vitro produced embryos have a lower viability. One might expect that this would be nicely compensated by transferring more embryos to each recipient, but this is not always the case. The second problem, polyspermy, provides another stumbling block in the use of in vitro produced embryos. The Petters' laboratory and many others have worked extensively on these problems and still they remain largely unresolved (Abeydeera 2001). Two recent papers provide us with some indication of the magnitude of the problem. In one case, only 20% of the oocytes were normally fertilized, the problem being a combination of low fertilization rate (60%) and high polyspermy rate (67% of fertilized oocytes) (Marchal et al. 2002). In the second case, similar results were obtained. A slightly increased fertilization rate (70%) and a lower polyspermy rate were reported (53%) leaving one-third of the oocytes normally fertilized (Wu et al. 2002). With practice, it may be possible to separate unfertilized from fertilized oocytes, this leaves the polyspermic (50–60%) and normal oocytes that will be subjected to centrifugation (see below) and probably microinjected. At least twice as many embryos may have to be microinjected in order to produce enough embryos for transfer. With some sort of compensation for the predicted lower viability, these embryos can be transferred to recipients. This approach has a number of potential advantages: (1) not as many animals are needed because embryo donors are not required; (2) surgical recovery of embryos is not required. Disadvantages may include: (1) more embryos may have to be injected because of lower viability and decreased number of normally fertilized embryos;

(2) recipients may be suboptimal and so pregnancy rate might suffer. Since the advent of cloning, this technique can be combined with transgenesis by using in vitro matured oocytes and somatic cells that have been genetically modified. For a description of cloning techniques, see Chap. 4 Cloning the Laboratory Mouse by Nuclear Transfer (K. Eggan and R. Jaenisch) and Chap. 8 Nuclear Transfer in the Cow (W.A. Ritchie), in this manual.

9.2 Protocol for In Vitro Production of Pig Embryos

9.2.1 Media, Solutions and Reagents

There are two media systems currently available for use for the in vitro production of pig embryos: the standard system and the PPM (Purdue Porcine Media, developed by Rebecca Krisher) system. The details of media components for all media needed for each step in each system are given below.

Procedure	PPM media system	Standard media system
Maturation	PPMmat	TCM199
Fertilization	PPMfert Synthetic oviductal fluid (SOF) HEPES	mTBM SOF HEPES
Sperm wash	D-PBS Percoll (45% and 90%)	Sperm wash medium D-PBS Percoll (45% and 90%)
In vitro culture	PPM1, PPM2	NCSU23

9.2.1.1
TCM199 Oocyte Maturation Medium

Reagent	Defined	Undefined
TCM199	9.5 ml	8.5
Polyvinyl alcohol	0.01 g	–
PFF[a]	–	1.0 ml
Glucose (3.05 mM; 0.5496 g/10 ml TCM199)	100 μl	100 μl
Pyruvate (0.91 mM; 0.05 g/5 ml TCM199)	100 μl	100 μl
PSA[b]	100 μl	100 μl
Cysteine (0.5 mM; 0.05 g/5 ml TCM199)	100 μl	100 μl
pLH (porcine luteinizing hormone; 0.01 U/ml)	50 μl	50 μl
pFSH (porcine follicle stimulating hormone; 0.01 U/ml)	50 μl	50 μl
EGF (epidermal growth factor; 50 ng/ml; 0.1 mg/10 ml D-PBS)	10 μl	10 μl

[a] Porcine follicular fluid

[b] Penicillin/streptomycin/amphotercin, an antibiotic antimycotic

Add reagents to TCM199 the morning of oocyte collection. Filter, make wells and place wells and medium in the incubator.

9.2.1.2
In Vitro Fertilization Medium (mTBM)

- mTBM Stock (modified Tris-buffered medium)

Reagent	mM	g/100 ml
NaCl	113.1	0.6611
KCl	3.0	0.0224
$CaCl_2 \cdot 2H_2O$	7.5	0.1102
Tris	20.0	0.2423
Glucose	11.0	0.1982
Na-pyruvate	5.0	0.0550

Filter. After preparation the pH should be about 10. Do not adjust. Store at 4 °C. Good for 2 weeks.

- mTBM working solution
  - Combine 9.9 ml mTBM stock, 0.004 g caffeine (2 mM), 100 μl PSA (penicillin/streptomycin/amphotercin, an antibiotic antimycotic) and 0.02 g fraction V BSA. Filter and use.

9.2.1.3
Sperm Wash Medium

- D-PBS w/o $CaCl_2$
  - Make D-PBS according to directions on packet, omitting the $CaCl_2$. Make 1 l and adjust pH to 7.2–7.4. Filter. Store at 4 °C. Good for 3 months.
- Sperm wash medium working solution
  - Add 0.05 g fraction V BSA and 500 μl PSA to 50 ml D-PBS.

9.2.1.4
Percoll

To make 90% Percoll solution, combine 45 ml Percoll (Sigma) with 5 ml 10 × stock of modified sperm TL.

- 10x stock modified sperm TL

Chemical	g/100 ml
KCl	0.230
NaH_2PO_4	0.035
NaCl	4.675
HEPES	2.380

Adjust pH to 7.3

To this 50 ml of 90% Percoll, add:

Chemical	Per 50 ml
$CaCl_2$	14.5 mg
$MgCl_2 \cdot 6H_2O$	4.0 mg
Lactic acid	0.184 ml
$NaHCO_3$	104.5 mg

9.2.1.5
Synthetic Oviductal Fluid (SOF) HEPES Medium (Liu and Foote 1995; Tervit et al. 1972; Gardner et al. 1994, 1997; Gandhi et al. 2000)

- Stock A (base) 10×; 3 months expiration

Reagent	mM	g/100 ml	g/200 ml	g/500 ml	g/1,000 ml
NaCl	99.7	5.827	11.654	29.135	58.270
KCl	7.16	0.534	1.068	2.670	5.340
KH_2PO_4	1.19	0.162	0.324	0.810	1.62
$MgCl_2 \cdot 6H_2O$	0.49	0.0996	0.1992	0.498	0.996
Na lactate (60% syrup)	3.30	0.616 (280.25 μl)	1.232 (560.25 μl)	3.08 (1,401.25 μl)	6.16 (2,802.5 μl)

- Stock D (calcium) 100 ×; 1 month expiration

Reagent	mM	g/10 ml	g/20 ml	g/50 ml
$CaCl_2 \cdot 2H_2O$	1.71	0.252	0.504	1.260

- Stock G (glucose) 10×; 3 months expiration

Reagent	mM	g/10 ml	g/50 ml	g/100 ml
Glucose	1.50	0.0270	0.135	0.270

- Stock HEPES (HEPES) 10×; 3 months expiration

Reagent	mM	g/50 ml	g/100 ml	g/200 ml
HEPES, acid	21	2.9795	5.958	11.916
Phenol red (0.5% solution)		(0.69 ml)	0.0069 g(1.38 ml)	(2.76 ml)

Place approximately 400 ml milliQ water in a 1,000 ml beaker with a stir bar. Mix solutions according to the following directions under the hood:

Stock solution	HEPES (1 l)
A (base)	100 ml
D ($CaCl_2$)	10 ml
G (glucose)	100 ml
NEAA	10 ml
EAA	20 ml
HEPES	84 ml

Place beaker on the stir plate. To the solution add:

Chemical	For 1 l
$NaHCO_3$	0.336 g
Na pyruvate	0.036 g
Glutamine	0.146 g
Fraction V BSA	1.0 g
Gentamycin	1,000 μl
Amphotercin (fungizone)	1,000 μl

Let the solution stir gently until BSA dissolves. This may take a while. Bring volume to approximately 900 ml. pH solution to 7.3–7.4. QS to 1 l. Filter, label and store at 2–8 °C.

Caution: pour the solution slowly down the side of the volumetric to avoid bubbling.

Essential and non essential amino acids are purchased from ICN (ICN Biochemicals, Aurora, OH);

- 1601149 essential amino acids, 50 ×
- 1681049 non essential amino acids, 100 ×

Amphotercin (Fungazone; # 15290–018) from Gibco (Rockville, MD)
Gentamycin (# G 1397) from Sigma (St. Louis, MO)

9.2.1.6
NCSU 23 (Petters and Wells 1993)

Component	g/l	mM
NaCl	6.35	108.7
KCl	0.356	4.78
KH_2PO_4	0.162	1.19
$MgSO_4 \cdot 7H_2O$	0.294	1.19
$NaHCO_3$	2.106	25.07
Glutamine	0.146	1.0
Glucose	1.0	5.55
Taurine	0.876	7.00
Hypotaurine	0.545	5.00
$CaCl_2 \cdot 2H_2O$	0.189	1.70
BSA fractionV (Sigma 3311)	4.0	
Phenol Red, 0.5% (Sigma P0290)	1.0 ml	
Penicillin/Streptomycin (Sigma P0781)	10.0 ml	

Note: All chemicals should be of embryo culture quality. Place components 1–9 in an acid-washed or oven-baked 1 l volumetric flask and add embryo culture quality water to about one-half the volume of the round base of the flask. Swirl gently to get all of the components into solution.

- In a separate 50 ml beaker that has been properly cleaned, completely dissolve the $CaCl_2$ $2H_2O$ in 20 ml embryo culture grade water. After components 1–9 are in solution add the $CaCl_2$ $2H_2O$ slowly and drop-wise to the hand-swirled flask (if a white precipitate forms, start over from scratch). Once components 1–10 are in solution, add the BSA to the medium.
- Allow the BSA to float on top of the medium and slowly dissolve into the solution. Once components 1–11 are in solution, add the phenol red and antibiotics and bring volume to exactly 1 l with embryo culture grade H_2O. Filter-sterilize the medium through a 0.2 μm vacuum filter flask (Nalgene 154–0020).

9.2.1.7
PPM Media System

Purdue Porcine Media for porcine maturation, fertilization and embryo culture (R.L. Krisher, Purdue University).

- PPM Base (10 × stock, 1 month expiration; start with 50 ml milliQ H_2O, add reagents, QS to 100 ml)

Reagent	MW	Concentration in final working solution (mM)	g/100 ml for 10 × stock
$NaCl_2$	58.44	100	5.8440
KCl	74.55	5.0	0.3728
KH_2PO_4	136.1	0.50	0.0681
$CaCl_2$	111.0	1.70	0.1887
$MgSO_4 \cdot 7H_2O$	246.5	1.19	0.2933
Alanine	89.09	0.50	0.0446
Glycine	75.07	2.0	0.1501
Phenol red	354.4	0.014 (0.0005%)	0.0050

- 10 × Stock solutions

Stock	MW	Concentration in final working solution (mM)	g/10 ml for 10 × stock	Expiration
Bicarbonate ($NaHCO_3$)	84.01	25	0.21	1 week
Glucose	180. 2	2.0	0.0360	1 month
L-Lactate	90.08	6.0	0.0541	1 month
Pyruvate	110.0	0.2	0.0022	1 week
Taurine	125.1	5.0	0.0626	1 month
Hypotaurine	109.1	5.0	0.0546	1 week

- 100 × Stock solutions

Stock	MW	Concentration in final working solution (mM)	g/10 ml for 100 × stock	Expiration
Glutamine (GLN)	146.15	1.0	0.1461	1 week
Cysteamine	77.14	0.5	0.0386	1 week
Cysteine	175.64	0.57	0.10	1 week
Citrate	294.1	0.5	0.1471	1 week

- PPM Media

  Note: All volumes given in the following table are in microliters unless otherwise stated.

Stock	PPMmat.	PPMfert.	PPM1	PPM2
MilliQ water	1,750	1,220	1,600	1,303
Base	500	500	500	500
Bicarbonate	500 (25 mM)	880 (44 mM)	500 (25 mM)	500 (25 mM)
Glucose	500 (2 mM)	100 (0.4 mM)	100 (0.4 mM)	1,500 (6 mM)
L-Lactate	500 (6 mM)	500 (6 mM)	500 (6 mM)	167 (2.04 mM)
Pyruvate	25 (0.01 mM)	500 (0.2 mM)	500 (0.2 mM)	250 (0.1 mM)
Glutamine	50 (1 mM)	50 (1 mM)	50 (1 mM)	75 (1.5 mM)
Taurine	500 (5 mM)	700 (7 mM)	700 (7 mM)	500 (5 mM)
Hypotaurine		500	500	
Cysteamine	50			
Cysteine	50			
Citrate	50			
Caffeine, 2.06 mM		0.002 g		
NEAA, 1×	50	50	50	50
EAA, 0.5×	50			50
Vitamins, 1×	50			50
Hyaluronate, 0.5 mg/ml	250			
BSA, 2 mg/ml fert, 4 mg/ml culture		0.010 g fatty acid free	0.020 g pentex	0.020 g pentex
LH, 0.01 U/ml	25			
FSH, 0.01 U/ml	25			
EGF, 50 ng/ml	25			

Insulin/ transferrin/ selenium (ITS), 1× (I=5 μg/ml, T=5 μg/ml, S=5 ng/ml)	50			50
GII, 100 ng/ml				5
Total volume	5 ml	5 ml	5 ml	5 ml

9.2.2 Timeline

This detailed protocol, which is used in the Krisher laboratory, is based on that of (Brad et al. 2003), (Gandhi et al. 2001), and (Swain et al. 2001, 2002).

−4 h	Prepare maturation dishes and allow to equilibrate in incubator
0 h	Arrival of ovaries, collection of oocytes and incubation for maturation. Make mTBM fertilization medium working solution and incubate to equilibrate
+24 h	Prepare mTBM fertilization drops and sperm wash medium, or PPMfert medium and drops. Also prepare SOF HEPES and D-PBS
+38 h	Make 45% and 90% Percoll
+40–42 h	Thaw sperm, separate and count. Denude oocytes. Co-incubate sperm and oocytes for 5 h
+42 h	Prepare zygote culture dishes and incubate
+45–47 h	Place fertilized zygotes into culture
+52–54 h	Mount and fix zygotes on slides to check for fertilization, if desired
+189–191 h	Score for cleavage and development (144 h post culture)

9.2.3 Oocyte Collection and In Vitro Maturation (Day 1)

- Reagents required:
  - SOF HEPES, (HEPES), medium
  - maturation medium (TCM199 or PPMmat.)
  - Sigma mineral oil, embryo-tested

On the morning ovaries are expected, make your working media for the day. If the PPM system is being used, media should be prepared the evening before use to allow sufficient equilibration time. Make sure that 1 l saline is placed in a warm oven. Aliquot 100 ml HEPES medium into a bottle and place it in the dry oven to warm as well. Make the appropriate maturation medium (TCM199 or PPMmat). Sigma embryo-tested mineral oil should be used throughout this protocol. Prepare a Nunc four-well plate by placing 500 μl of the appropriate maturation medium into each well and cover with about 400 μl mineral oil. One well is used for washing. Place the Nunc plate in the 5% CO_2 in air incubator at 39 °C to equilibrate, preferably for 4 h. Both maturation and fertilization are carried out at 39 °C.

1. Set up for aspiration. When the ovaries arrive, take the temperature of the ovaries and the saline. Pour the ovaries into a strainer and wash them twice with the warmed saline, then place them in a beaker and fill the beaker with fresh warm saline so that the ovaries are covered. Do this quickly so they do not get cold. Place the beakers on a slide warmer by the aspiration pump.

2. Aspirate follicles ranging in size from 3 to 8 mm in diameter on the ovary. Pierce the surface of the ovary just outside the visible boundary of the follicle. Go into the follicle under the surface of the ovary, and collect the fluid. While aspirating the fluid, use the needle to scrape the follicle walls to ensure the oocyte is recovered. To avoid loss of fluid, keep the bevel of the needle pointed down and away from you. Collect the fluid in 50 ml tubes.

3. All work with oocytes and embryos in this protocol should be performed as quickly as possible. All dishes containing oocytes or embryos should be worked with on slide warmers where possible, and time out of the incubator for all oocytes, embryos and/or dishes should be minimized. Opening of the incubator door should always be minimized, in regard to number of times opened, length of time open, and width of opening. Red light should be used whenever oocytes or embryos are out of the incubator, because natural light wavelengths can be harmful to oocyte and embryo development. Extended exposure to adverse conditions will result in reduced embryo viability or death.

4. Set up dishes to be used in the search for oocytes: 1× Falcon 1007 dish, 2× Falcon 1008 dishes, several grid dishes (number of tubes of follicular fluid +1), a Drummond pipette (Drummond Scientific, cat. no. 3-000-105) with tips, and a sterile disposable transfer pipette. Draw 7–8 lines on the bottom of the 1007 dish. Fill the 1007 and 1008 dishes with HEPES medium – you can either pour or pipette. Aspirate, or sweep, the

pellet at the bottom of one aspiration tube using the transfer pipette. Pull sediment (cellular debris and oocyte cumulus complexes) off the bottom of the 50 ml tube into the pipette, using a circular motion. Place the sediment in a grid dish and add HEPES medium. Swirl to mix. With a new Drummond tip, search the grid dish for oocytes. The magnification should be such that one square of the grid dish fits in the field of view. Perform some oocyte selection at this time. Ultimately, only oocytes completely covered with cumulus cells that are tightly compacted, and with an even cytoplasm (not pycnotic) will be selected. Also pay attention to the size of the oocyte cytoplasm. If the cytoplasm appears smaller than normal, do not select it. Do not select oocytes with expanded cumulus masses, or with no or incomplete cumulus layers. At this time only a brief look should be given each oocyte. Most oocytes will be removed from the grid dish at this time unless they are obviously poor (such as denuded or expanded). Use the Drummond pipette to pick the oocytes out of the grid dish and move them into the 1007 dish. Search the entire grid dish for oocytes. After the first search, swirl the dish again and search again. Repeat this procedure for each tube. Do a second sweep on all the tubes combined (all second sweeps in one grid dish) when finished with all the first sweeps.

5. When all the oocytes have been collected in the 1007 dish, change the Drummond tip (sterilize the stylet of the pipette with alcohol). Increase the magnification and examine each oocyte line by line in the 1007 dish for the selection criteria described above. Pick up selected oocytes that meet the criteria and place them into the first 1008 dish. When all oocytes have been examined and either selected or rejected, wash the selected oocytes through the remaining 1008 dish. If oocytes that should not be selected become apparent during the washes, they should be discarded at this time.

6. Remove the Nunc four-well dish with maturation medium from the incubator. Place oocytes from the final 1008 wash dish into the wash well, swirl and then distribute to the culture wells, 50 oocytes per 500 μl Nunc well. Place the maturation wells back in the incubator. Oocytes are matured in this medium for 40–42 h in 6% CO_2 in air at 39 °C.

7. On this day be sure to record in your notebook:

   - when the ovaries arrived
   - temperature of ovaries on arrival
   - temperature of saline
   - when aspiration began and ended

- number of ovaries you aspirated
- when the oocytes went into maturation medium
- how many oocytes were recovered
- maturation treatments/medium used

8. Also on this day make fertilization medium working solution and place in the incubator to equilibrate if using the mTBM fertilization medium. Do this in the afternoon.

9.2.3.1
In Vitro Maturation (Day 2)

- Reagents required:
  - Fertilization medium (mTBM or PPMfert)
  - Sperm wash medium
  - Percoll
  - SOF HEPES, (HEPES)
  - D-PBS

1. Oocytes remain in the maturation wells today.
2. In the afternoon, make fertilization drops if using mTBM. Drops should contain 50 μl medium; 20 oocytes will be fertilized per drop. Also make a wash dish with 3 × 250 μl drops. When making drops, place one-half the volume of the drop in the Falcon 1007 dish, cover with 10 ml mineral oil, then add the remaining drop volume. Place in the incubator (5% CO_2 in air) to equilibrate. In the PPM system, medium and drops are made on this day.
3. Also in the afternoon, make 50 ml sperm wash working solution in a bottle and place in a warm oven if using mTBM. Also prepare and warm 10 ml each SOF HEPES and D-PBS.

9.2.4
In Vitro Fertilization

- Reagents required:
  - mTBM or PPMfert
  - 90 and 45% Percoll
  - D-PBS sperm wash working solution
  - D-PBS
  - SOF HEPES, (HEPES)

1. In the morning of the day of fertilization, make sure the slide warmers and water bath are turned on. Make 45% (2 ml; 1 ml 90% +1 ml D-PBS with no BSA) and 90% Percoll (1.5 ml) solutions and place in a warm oven. Place the centrifuge buckets in a warm oven as well.
2. The first task is to denude the oocytes of cumulus mass cells. If you are experienced, you can do this during the first Percoll spin, but it must be accomplished in less than 20 min. If you are relatively inexperienced, denude the oocytes first then prepare the sperm. The oocytes should be fertilized 40–42 h post maturation. There are two methods for denuding the oocytes. For most uses, vortexing is the easiest. For nuclear transfer, use the Pipetman. In this case, minimize the exposure of oocytes to hyaluronidase as much as possible.
   (a) Vortexing. Place 100 μl warmed HEPES into a 1.5 ml microfuge tube. Add 1 μl hyaluronidase solution. Get out 2 × 1008 dishes, fill one with HEPES. Get out the maturation wells, remove the oocytes from the wells using the Drummond and place them in the microfuge tube. Vortex the tube for 3 min. Rinse the tube with HEPES and place the contents in the empty 1008 dish. Repeat twice more. The denuded oocytes will now be in the 1008 dish.
   (b) Pipetting. Prepare Falcon 3037 dishes by adding 4 ml HEPES to the outer moat, and 1 ml to the inner well. Add 10 μl hyaluronidase solution to the inner well. Take the maturation dishes out of the incubator. Using a cleaned Drummond, remove the oocytes from the maturation drops and transfer them into the inner well. Do only as many as you can quickly process at a time. After placing them into the hyaluronidase well, use a Pipetman set to 100–200 μl and quickly draw the oocytes in and out several times. Once a few oocytes are completely denuded, move them out of the hyaluronidase solution into the HEPES in the moat with the Drummond. Go back and denude a few more with the Pipetman. Repeat this process until all the oocytes in that group are denuded and in the moat. Using the Drummond, remove the group from the moat into a new 1008 dish with HEPES medium. Repeat until all oocytes are denuded.
3. Swirl the 1008 dish with the newly denuded oocytes to rinse well. All hyaluronidase must be rinsed away. After swirling, move the oocytes into the clean HEPES in the second 1008 dish in as little medium as possible. Swirl again. Collect into a group.
4. Move the oocytes into the fertilization medium wash drops. Wash them vigorously through all three wash drops by pulling oocytes and medium up and down rapidly in the Drummond. When all the

oocytes are in the final fertilization wash drop, get the previously prepared fertilization drops out of the incubator. From the wash dish, move the oocytes into the fertilization drops with as little medium as possible (20 oocytes per drop).

5. Once the oocytes are denuded and in the fertilization drops, you are ready to begin preparing the sperm. Make a Percoll gradient by gently layering 1.5 ml 45% Percoll on top of the 1.5 ml 90% Percoll in the existing 15 ml blue cap centrifuge tube. Frozen sperm: to thaw the sperm, drop two frozen pellets into 2 ml warmed HEPES in a small snap cap tube, mix gently once thawed. Layer this 2 ml gently onto the Percoll gradient. Fresh/chilled sperm: Remove 1 ml extended chilled semen from the large 50 ml blue capped tube in the chiller (mix before removing) and warm in a small snap cap tube. Place 500 μl fresh/chilled semen onto the Percoll gradient. Spin the Percoll for 20 min at 700 g (2,100 rpm). Remove the supernatant down to the pellet. Add 5 ml sperm wash medium (mTBM system) or D-PBS (PPM system) to the pellet and flick it gently with your finger to mix the pellet well. Spin for 5 min at 700 g. Again, remove the supernatant down to the pellet, and repeat the rinse. After the final spin, add 100 μl fertilization medium to the pellet and mix gently by hand.
6. Prepare the dilution tubes for sperm counting. Fill one 1.5 ml microfuge tube with 900 μl water and label 1,000 ×, and one with 990 μl water and label 100 ×. Add 10 μl sperm solution to the 100 × tube, mix well. Take 100 μl of this solution and add it to the 1,000 × tube. Mix well. Place the sperm solution, as well as the remaining fertilization medium, in the incubator for 1 h.
7. Count the sperm and dilute. Place 10 μl 1,000 × dilution on each side of the haemocytometer. Count the four large grid boxes on each side of the haemocytometer (a total of eight boxes). Divide the total sperm counted by eight. This is the average count per square, or x. Multiply by 10 (each square of the haemocytometer has a volume of 0.1 mm^3) =10 × sperm/μl. Multiply by 1,000 to account for the dilution factor, and multiply by 1,000 again to get sperm/ml. Thus, average count × 107 is sperm/ml. The final sperm concentration in the fertilization drop should be 500,000 sperm/ml. Thus, we need to add 50,000 sperm in 50 μl to give a total of 50,000 sperm in the 100 μl drop. 50,000 sperm in 50 μl = 1 million sperm/ml. This is the correct sperm dilution. To calculate how much of the sperm pellet and how much of the fertilization medium to mix to make a 1 million

sperm/ml suspension, use the following calculation:

$$\frac{\text{Average count} \times 10^7}{1,000\mu l} = \frac{1 \times 10^6}{x\mu l}$$

Solve for x. This is the amount of the sperm pellet to add. Subtract this number from 1,000 to get the amount of fert medium to add. Examples:

(a) If your total count of eight boxes was 12 sperm: $12 \div 8 = 1.5$, so there are 1.5×10^7 sperm/ml in the sperm pellet tube.

$$\frac{1.5 \times 10^7 \text{sperm/ml}}{1,0000\mu l} = \frac{1 \times 10^6 \text{sperm/ml}}{x\mu l} \tag{9.1}$$

$x = 66.7$ μl to make the final sperm suspension, add 66.7 μl of the sperm pellet to 933.3 μl fert medium (1,000–66.7) in a microfuge tube.

(b) If the total count of eight boxes was 168: $168 \div 8 = 21$

$$\frac{2.1 \times 10^7 \text{sperm/ml}}{1,0000\,\mu l} = \frac{1 \times 10^6 \text{sperm/ml}}{x\mu l} \tag{9.2}$$

$x = 4.8$ μl to make the final sperm suspension, add 4.8 μl of the sperm pellet to 995.2 μl fert medium (1,000–4.8) in a microfuge tube.

In the calculation, if you need to inseminate with a final sperm concentration other than 500,000/ml, substitute 1×10^6 sperm/ml in the equation for the following:
250,000, use 0.5×10^6
500,000, use 1×10^6
1 million, use 2×10^6
2 million, use 4×10^6

8. Remove the fertilization drops containing oocytes from the incubator. Place them on the slide warmer. Using a Pipetman, add 50 μl of the sperm dilution. Check to be sure that you have added sperm to all drops containing oocytes and put the dishes back into the incubator. Clean up the area. Record in your notebook the time the oocytes were moved to fertilization drops, the counts, the dilution made, and the time the sperm was added to the oocytes. Co-incubate oocytes and sperm for 5 h.

9.2.5 In Vitro Culture

- Reagents required:
  - NCSU23 or PPMI [followed by PPM2 at 72 hpi (hours post insemination)]

1. The presumptive zygotes should be placed into culture 5 hpi. This is the same day as fertilization. Embryos are cultured at 39 °C in 6% CO_2 and 10% O_2, balanced with N_2. On the day of fertilization and culture, you should make the appropriate culture medium (NCSU23 or PPM1) immediately after fertilization. Make drops (25 μl + 10 ml oil +25 μl) in 1007 dishes as well as three 250 μl drops (125 μl+ oil +125 μl) in a 1007 dish for washing. Place in the incubator to equilibrate. The drops should equilibrate at least 3 h before adding the zygotes.
2. Remove the fertilization and the culture medium wash dish from the culture incubator. Wash the fertilized zygotes well through all three wash drops of the culture medium. (If you are mounting embryos on a slide to check for fertilization, place the zygotes to be mounted in the wash drops back into the incubator. At 12 hpi, mount them on a slide.)
3. Remove the culture drops from the incubator. Pick up ten embryos in as little medium as possible and place into the final culture drop. Place the culture dish back into the incubator. Record the culture medium used, and the time the zygotes were placed into the culture drops.
4. If NCSU23 is used, nothing else is done until day 6. If PPM1/2 is used, embryos are moved from PPM1 to PPM2 at 72 hpi. On day 3 (72 hpi), make PPM2 culture medium. Make culture drops in the morning (25 μl + oil) and allow to equilibrate for at least 3 h. Again, make wash drops in the culture plate as above. At 72 hpi, move the embryos quickly from the original culture drops to the wash drops, wash through three wash drops and distribute ten per 50 μl culture drop. Embryos must be washed thoroughly at this step. Move them in as little of the original medium as possible at each step. Place the new culture drops back into the incubator. Record the date and time when the embryos were moved. NCSU23 allows for continuous culture without intervention or changing of medium. PPM1/2 media are designed to be changed during the culture period. Since NCSU23

does as well or better for most applications, its ease of use argues for its inclusion in transgenic protocols with pigs.

5. At 144 hpi (day 6) post insemination, remove the embryos from the incubator and score each embryo for cleavage and development.

9.3 Protocols for In Vivo Production of Pig Embryos

9.3.1 Media, Solutions and Reagents

NCSU23. See Sect. 9.2.1.6 for components.

9.3.2 Timeline

−0 h	Receipt of gilts. PMSG dosage (4:00 pm)
+72 h	hCG dosage (4:00 pm)
+114 h	First natural mating of gilts (10:00 am)
+119 h	Second natural mating of gilts (3:00 pm)
+121 h	Withdraw food supply from gilts prior to surgery (5:00 pm)
+138–143 h	Surgery for recovery of embryos (10:00 am–3:00 pm)
+139–141 h	Microinjection of embryos (11:00 am–4:00 pm)
+143–145 h	Transfer of microinjected embryos (3:00–5:00 pm)

9.3.3 Synchronization and Superovulation of Embryo Donors

Sufficient embryo donors and recipients are required to efficiently produce transgenic pigs. In our hands, a minimum of five prepuberal gilts (young female pigs) are required to provide sufficient embryos on average to allow for one successful embryo transfer. Gilts (150–160 days of age) were shipped to our farm on the day superovulation was to begin (Sommer et al. 2002). Shipping stress may enhance the success of superovulation. Pregnant mare's serum gonadotrophin (PMSG, 1,500 IU) was injected by intramuscular injection (im) at 4:00 pm on the day of arrival. Human chorionic gonadotrophin (hCG, 500 IU, im) was given 72 h after the PMSG injection. In the United States these drugs are not approved for general use in pigs and must be cleared through Federal and State authorities. Special permission will be required. The gilts were then mated by natural service or artificial insemination at 42 and 47 h after the hCG injection. The synchronization of gilts is an advantage in this procedure since the gilts are bred by artificial insemination on a timed basis, irrespective of oestrus behavior. Oestrus is the period during which the female is receptive to mating by the male. When properly stimulated, she assumes a characteristic "standing oestrus" pose indicating receptivity. Oestrus in the pig lasts for 3 days. Natural mating may result in ovulation occurring during a fairly broad time frame during oestrus. Superovulation with gonadotrophins in prepubertal females "sets" the system so that we can better predict the time of ovulation. Two inseminations are given to be sure of fertilization. Because oestrous behavior is variable, inseminations are given regardless of behavior. Exceptionally good results have been obtained with this system of reproductive management. Our experience indicates that virtually every donor ovulates using this protocol and will produce fertilized embryos if inseminated. Without synchronization, one would not know when to breed or when ovulation was occurring during the 3 day oestrous period. Synchronization also allows a greater number of embryo donors to be handled on the same day, which may be important for obtaining sufficient embryos for transfer. Surgical collection should start on the morning after insemination. Most of the embryo donors can be expected to have more than 20 embryos. Thus, a group of five donors should yield on average 100 embryos for microinjection. Experiments should be avoided when the ambient temperature is very high, due to the effects of high temperature on fertility. These effects become evident in males, which exhibit a pronounced sterility, and also in females, which undergo a behavioral depression.

9.3.4 Embryo Recovery

Embryo recovery in the pig requires surgery and may be intimidating to most laboratories. If this is the case, collaboration with laboratory animal experts may be warranted. The strategy to be used is to enter the body cavity, exteriorize the oviducts, and flush them with embryo culture medium. Details on the anaesthesia and the surgery are best described by surgical experts (Swindle 1998) as a number of different protocols are possible.

- Reagents required:
  - NCSU 23

9.3.4.1
Preparation for surgery (from (Swindle 1998))

1. Pigs are fasted from solid food for at least 6–8 h
2. When going into surgery, the pig is draped to prevent hypothermia
3. Monitoring may be accomplished by noting respiration and using a pulse oximeter
4. An injectable or inhalation anaesthesia may be used
5. Example anaesthesia protocol ((Swindle 1998), *General Surgery Protocol*, p 53)

Induction	
Ketamine	33 mg/kg im
Acepromazine	1.1 mg/kg im
Atropine	0.05 mg/kg im
Maintenance	
Isofluorane	1.5%–2% in oxygen or 0.5%–1.5% in nitrous oxide; oxygen 2:1

6. One person (the anesthesiologist) must monitor the vital signs of the patient during the surgery. The time required for surgery must be minimized; no more that 30 min from the time of induction of anaesthesia to removal from isofluorane is a good goal to try to achieve.

9.3.4.2
Surgical Technique for Skin Opening and Closure

These techniques are discussed in Swindle's books (Swindle 1983, 1998). The choice of instruments (Swindle 1983) is important, as are the suture and needles. Our own suture methods have evolved over time. The reason for this is that the abdominal incision in a large pig is subject to a lot of abdominal pressure, and scratching may damage the suture line on the skin. On the internal suture line of peritoneum and linea alba taken up together, we insert the needle about 12.5 mm on each side of the incision using synthetic suture, thus turning the edges of the suture line slightly up. In addition, each stitch is "locked" by turning the suture loop and pulling the needle through it. Consequently, there is little chance that any of these pigs will ever develop a hernia. The skin incision could also be a problem. If scratched, the suture may dehisce. We use a "box" stitch with cat gut that is absorbable and thus does not have to be removed later. After appropriate withdrawal times, embryo donors went on to slaughter and embryo recipients were held for anticipation of pregnancy and parturition about 4 months later.

9.3.4.3 Embryo Collection (Protocol from Petters' laboratory)

Our protocol was originally developed at North Carolina State University in the 1960s by L. C. Ulberg – a pioneer in embryo transfer research.

Blunt dissection (using only fingers) is used through the substantial fat layer found between the skin and the connective tissue overlying the abdominal cavity. A ∼ 5–7.5 cm incision is made in the linea alba (midventral connective tissue between abdominal muscles) to reveal the abdominal region. The abdominal cavity is entered by "poking" with a finger to pierce the peritoneum and then enlarging the hole to the size of the incision in the linea alba. One reproductive tract is exteriorized at a time, and the oviducts flushed with sterile medium (see below). After both sides have been flushed, the reproductive tracts are carefully returned to the abdominal cavity and the incisions closed immediately unless the animal is to be held for use as an embryo recipient (see below).

For recovery of one cell embryos, a bent glass cannula (sized to fit the embryo donors being used – a range should be available when first starting), flared at one end, is placed in the upper end of the oviduct, the infundibulum, which is a large membraneous structure that enshrouds the ovary. The cannula can be carefully clamped, but we have found much better results if a surgical assistant holds the cannula in place between thumb and forefinger and directs the end of the cannula to the collection dish. The surgeon now uses syringe and blunted needle (18 g) to cannulate the uterine tubule junction (UTJ) and complete a retrograde flush of the oviduct using 15–20 ml embryo culture medium (NCSU-23). A team of at least three people are required for the surgical activity (surgeon and assistant + anesthesiologist). This team of three continues to isolate embryos in the surgery while a second team of at least two people will handle and microinject the embryos in a separate room very close to the surgery. As soon as embryos are obtained, they are taken for processing and microinjection. Close coordination of recovery, microinjection and transfer became a key goal of the teamwork approach.

9.4 Production of Transgenic Pigs

9.4.1 Centrifugation and Microinjection of Embryos

Pig embryos are sensitive to low temperature, so room temperature (assumed to be 25 °C) is kept a few degrees higher if possible (25–38 °C). Even room temperature is a risk to pig embryos. Warming trays and incubators are to ensure that the embryos never experience temperature below 38 °C except for short periods of time – even the microfuge is placed inside an incubator, although the incubator must be monitored for heat build up. During microinjection, a heated stage is used to maintain embryo viability. Water-jacketed tissue-culture incubators are used. Cheaper, air-jacketed, thermal incubators must never be used because the temperature may reach levels higher than desired. Temperature calibration of incubators and other such devices must be determined and verified by the average of three calibrated thermometers. Never autoclave anything that will come into contact with embryos because of steam cleaning residues – instead use high dry heat (glassware), boiling water (plastic tubing) and filter (media) sterilization.

Handling of embryos is kept to a minimum and must always be carried out in an efficient manner. NCSU-23 is used at all times and must be gassed with 5% CO_2. Whatever medium is used, switching to different media during the course of the experiment is discouraged, as is the use of oil overlays (except during microinjection and fertilization). Embryos are recovered from the oviductal flush in gridded plastic plates that can easily hold 15–25 ml fluid that allow for the dispersal of debris. The embryos have a very dark, dense cytoplasm and are easily visible in the plate of flushing medium. Embryos are collected using a mouth-controlled Pasteur pipette (pulled, and broken or cut to a working size tip). Embryos are washed a number of times and quickly placed in the incubator prior to centrifugation and microinjection.

Centrifugation allows visualization of the pronuclei (Wall et al. 1985). Embryos are placed in a standard microcentrifuge tube and centrifuged for a few minutes. Each laboratory will have to titrate this method to find the time that works best for visualizing the pronuclei. Consult the paper of (Wall et al. 1985) for more details on the technique, which is surprisingly easy to master.

Standard microinjection procedures (Polites and Pinkert 1994) using a heated stage can be used in the pig once the pronuclei are visible after centrifugation. While microinjection procedures may be standard, we have found two problems that may occur and which individual laboratories may have to grapple with, at least at some level. These are: making and filling the injection pipette, and achieving and maintaining adequate flow for microinjection. We have found that injection pipettes with internal capillaries fill best and whereas these used to be custom-made in the lab, they can now be purchased. Flow can be a problem because injection pipettes may have such a small opening that sufficient pressure cannot be generated to cause fluid to flow out. Hydrofluoric acid (HF) can be used to open the injection pipette once the pipette is loaded and attached to the microinjection setup.

Note: HF is a potentially harmful chemical. Consult with your Safety Office regarding handling, use and disposal.

Procedure for opening the tip of the microinjection pipette: A standard, plastic culture dish, such as a Falcon 3002, is used to hold the HF microdroplets. First, a 5 μl drop of HF is placed in the dish. Smaller droplets are made around the first drop. These droplets, in turn, are used to open the tip of the needle. Embryo culture quality mineral oil is used to overlay the HF droplets. The tip of the DNA-loaded needle is lowered into the oil and placed near one of the smaller droplets. The pressure to the injecting needle is then increased while the tip of the needle is directed into the HF droplet. Once flow is observed, the tip of the needle is moved out of the HF droplet and a small pool of DNA buffer is allowed to form around the tip of the needle as the flow dynamics of the needle are assessed by the operator. It may be necessary to repeat this procedure to get adequate flow.

This method has worked very well for us and was used in the production of our rhodopsin transgenic pigs. In our experience, survival after microinjection can be very high.

9.4.2 Embryo Recipient Selection and Embryo Transfer

Select small pigs and pigs with badly conformed legs for the first set of surgeries. These pigs will not be used as recipients. The best pigs are saved for the last surgeries and the best one or two of these (normal ovarian morphology, ovulation sites and no cysts), normal embryo number (10–25) and morphology (one-cell) are chosen as recipients for the microinjected embryos.

A total of 40–60 microinjected embryos are transferred to each recipient female. Either 20–30 can be transferred to each side or all embryos can be transferred to one side because the embryos will be transported and spaced evenly throughout both uterine horns. Survival at microinjection is not assessed. Instead, all microinjected embryos are speedily transferred to a recipient female. The embryos are carefully loaded into a 10 μl micropipette attached to a positive displacement micropipette by flexible tubing. The positive displacement pipette is an advantage because maximum control over flow is achieved and accidental problems such as temperature-induced volume changes are minimized. Using a dissecting microscope, load the micropipette. This can be a challenge because it must be held perpendicular to the bottom of the dish and the embryos and air bubbles must be loaded within the displacement of the micropipette (Fig. 9.1). A hole is made in the upper end of the oviduct (precise location is important) with

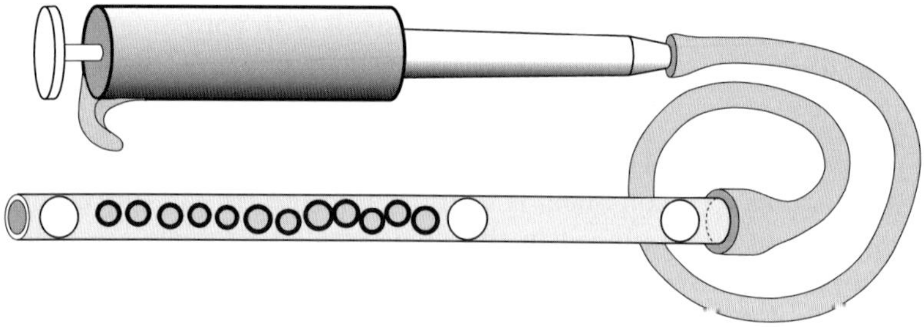

Fig. 9.1. Loaded Micropipette. Embryo transfer pipette attached by flexible tubing to positive displacement pipetter. (Drawn by Jeff Sommer)

a blunt needle, and the embryos carefully deposited in the oviduct in a minimal volume of medium, by first inserting the micropipette using the naked eye. Air bubbles in the capillary tube serve as convenient markers indicating when the embryos have entered the oviduct. Many laboratories use plastic tubing attached to a 1 ml syringe to instill the embryos directly into the upper end of the oviducts. Some possible problems include losing embryos (and perhaps not realizing it – this is especially a problem for the syringe method) and not placing the embryos in the proper location (not in the upper end of the oviduct, for instance). Generally, however, embryo transfer in the pig is routine and has a very high success rate.

9.4.3 Methods for Animal Identification

As with other transgenic animals, identification systems must be accurate and not likely to be lost or confused during the animal's lifetime. Founder animals must be screened with Southern blots that demonstrate unique flanking regions. This is absolutely required. Subsequently, PCR screens can be used and reconfirmed with Southern blots if required. When the pig is young, the tail is normally docked. This is the tissue used for DNA analysis. In pigs, tattoos are possible but are often difficult to read due to skin pigmentation and changes in size with growth. Ear notches and ear tags are used to identify pigs. In addition, cage cards with appropriate information are prominently displayed.

Acknowledgements. Contributors to this work include: Jeff Sommer, Bruce Collins, Curtis Alexander, Kevin Wells.

References

Abeydeera LR (2001) In vitro fertilization and embryo development in pigs. Reproduction 58 [Suppl]:159–173

Boquest AC, Grupen CG, Harrison SJ, McIlfatrick SM, Ashman RJ, d' Apice AJ, Nottle MB (2002) Production of cloned pigs from cultured fetal fibroblast cells. Biol Reprod 66:1283–1287

Brad AM, Bormann CL, Swain JE, Durkin RE, Johnson AE, Clifford AL, Krisher RL (2003) Glutathione and adenosine triphosphate content of in vivo and in vitro matured porcine oocytes. Mol Reprod Dev 64:492–498

Gandhi AP, Lane M, Gardner DK, Krisher RL (2000) A single medium supports development of bovine embryos throughout maturation, fertilization and culture. Human Reprod 15:395–401

Gandhi AP, Lane M, Gardner DK, Krisher RL (2001) Substrate utilization in porcine embryos cultured in NCSU23 and G1.2/G2.2 sequential culture media. Mol Reprod Dev 58:269–275

Gardner DK, Lane M, Spitzer A, Batt PA (1994) Enhanced rates of cleavage and development for sheep zygotes cultured to the blastocyst stage in vitro in the absence of serum and somatic cells, amino acids, vitamins, and culturing embryos in groups stimulate development. Biol Reprod 50:390–400

Gardner DK, Lane MW, Lane M (1997) Bovine blastocyst cell number is increased by culture with EDTA for the first 72 hours of development from the zygote. Theriogenology 47:278

Liu Z, Foote RH (1995) Development of bovine embryos in KSOM with added superoxide dismutase and taurine and with five and twenty percent O_2. Biol Reprod 53:786–790

Marchal R, Pelaez J, Terqui M, Mermillod P (2002) Effect of sperm survival and ctc staining pattern on in vitro fertilization of porcine oocytes. Theriogenology 57:1917–1927

Petters RM, Wells KD (1993) Culture of pig embryos. J Reprod Fertil 48 [Suppl]:61–73

Phelps CJ, Koike C, Vaught TD, Boone J, Wells KD, Chen SH, Ball S, Specht SM, Pelejaeva IA, Monahan JA, Jobst PM, Sharm SB, Lamborn AE, Garst AS, Moore M, Demetris AJ, Rudert WA, Bottino R, Bertera S, Trucco M, Starzl TE, Dai Y, Ayares DL (2003) Production of alpha 1,3-galactosyltransferase-deficient pigs. Science 299:411–414

Polites HG, Pinkert CA (1994) 2DNA microinjection and transgenic animal production. In: C. Pinkert (ed) Transgenic animal technology – a laboratory handbook. Academic Press, New York, pp 15–68

Sommer JR, Collins EB, Neiding T, Rozeboom K, Wong F, Petters RM (2002) Conservation and regeneration of transgenic lines of swine by semen cryopreservation and artificial insemination. Lab Animal 31:25–31

Swain JE, Bormann CL, Clark SG, Walters EM, Wheeler MB, Krisher RL (2002) Use of energy substrates by various stage preimplantation pig embryos produced in vivo and in vitro. Reproduction 123:253–260

Swain JE, Bormann CL, Krisher RL (2001) Development and viability of in vitro derived porcine embryos cultured in NCSU23 and G1.2/G2.2 sequential medium. Theriogenology 56:459–469

Swindle MM (1983) Basic surgical exercises using swine. Praeger, New York

Swindle MM (1998) Surgery, anesthesia and experimental techniques in swine. Iowa State University Press, Ames, IA

Tervit HR, Whittingham DG, Rowson LEA (1972) Successful culture in vitro of sheep and cattle ova. J Reprod Fertil 30:493–497

Wall RJ, Pursel VG, Hammer RE, Brinster RL (1985) Development of porcine ova that were centrifuged to permit visualization of pronuclei and nuclei. Biol Reprod 32:645–651

Wu G-M, Sun Q-Y, Mao J, Lai L, McCauley TC, Park K-W, Prather RS, Didion BA, Day BN (2002) High developmental competence of pig oocytes after meiotic inhibition with a specific m-phase promoting factor kinase inhibitor, butyrolactone i. Biol Reprod 67:170–177

10 Production of Transgenic Birds Using Lentiviral Vectors

Benjamin B. Scott, Carlos Lois

10.1 Introduction

The generation of transgenic birds using lentiviral vectors is an efficient procedure that is relatively simple compared to the production of transgenic mice and other mammals. Most tools and reagents required are available in a typical molecular biology laboratory, and the procedure can be performed on a laboratory bench without special concerns for sterility. The method for avian transgenesis described here is similar to previous methods used to produce transgenic chickens by the microinjection of recombinant oncoretroviruses into the early embryo of freshly laid eggs (Bosselman et al. 1989). However, lentiviral transgenesis has a number of important advantages over previous methods. The primary advantage of using lentiviral vectors is that they are not subject to the silencing that oncoretroviral-based vectors undergo in the developing embryo. Previous studies of transgenic chickens and quails produced with oncoretroviral vectors reported low or undetectable levels of transgene expression (Mozdziak and Petitte 2004). In contrast, we and others have demonstrated that lentiviral vectors allow for high levels of transgene expression in quails and chickens (Scott BB, Lois C: Tissue-specific expression in transgenetic birds. PNAS in press). Additionally, we have shown that lentiviral vectors allow for tissue-specific expression in transgenic birds. Transgenesis using lentiviruses has been shown to be a versatile and powerful genetic tool in mice (Lois et al. 2002). In this chapter we describe the application of this tool to birds. Although this system has only been tested in chickens and quails, we anticipate that lentiviral transgenesis will be possible in a wide variety of avian species.

Benjamin B. Scott: Massachusetts Institute of Technology, E25–421, 77 Massachusetts Avenue, Cambridge, MA 02139, USA, e-mail: bbscott@mit.edu

Carlos Lois: Massachusetts Institute of Technology, E25–421, 77 Massachusetts Avenue, Cambridge, MA 02139, USA, e-mail: los@mit.edu

Principles and Practice
Mammalian and Avian Transgenesis – New Approaches
Shirley Pease and Carlos Lois (Eds.)

10.2 Overview of the Strategy

In birds, after fertilization of the oocyte by the sperm has occurred, the embryo starts developing inside the female's reproductive tract until the egg is laid a few hours later. During the migration of the fertilized embryo through the reproductive tract, two key processes occur simultaneously. First, the cells in the embryo proliferate rapidly in such a way that by the time the egg is laid (3–5 days postfertilization) the embryo will consists of 20,000–50,000 cells (depending on the species). Second, during migration the egg will acquire a calcerous shell. Therefore, once the embryo has completed its intrauterine migration, the female will lay a single embryo consisting of tens of thousand of cells, which is surrounded by an opaque shell. These special characteristics of avian development pose some obstacles to the generation of transgenic birds. First, the embryo cannot be easily observed with a microscope, because the shell prevents its visualization. Second, because the embryo consists of some 20,000–40,000 cells it will not be possible to deliver transgenes to 100% of the embryonic cells. Accordingly, the strategy for transgenesis in birds is based on the delivery of transgenes to a percentage of the cells of the founder embryo. When this embryo develops, it will become a mosaic animal (F0), in which a percentage of cells will contain the transgene. If some of the transgene-carrying cells in the F0 mosaic founder differentiate into sperm or oocytes, then when these mosaic founder animals breed they will pass the transgene to a percentage of their progeny (F1), which will become transgenic birds carrying a copy of the transgene in all cells of the body.

10.3 Design of Lentiviral Vectors for Transgene Expression

A general introduction to the design and construction of lentiviral vectors is given in Chap. 1 of this book, Generation of Transgenic Animals Using Lentiviral Vectors. However, there are some specific considerations for the design of vectors used to make transgenic birds.

10.3.1 Promoters

The appropriate choice of cis-regulatory elements used to drive transgene expression is critical to the successful generation of useful transgenic animals. A variety of enhancer/promoter combinations are available to drive transgene expression ubiquitously, or in specific tissues, or at different times during development. However, most well characterized promoters have been studied only in mice or other mammals. Some promoter/enhancer combinations can have quite different transcriptional activity in mice than in birds, which can lead to changes in tissue specificity or overall expression level from species to species. These differences in gene expression can often be difficult to predict, especially when the promoter is a viral sequence that shares no homology with regulatory elements in the avian genome, such as the immediate early promoter from the human cytomegalovirus (CMV). In mice, the CMV promoter is commonly used to produce animals with strong transgene expression in most tissues (Schmidt et al. 1990; Furth et al. 1991). In contrast, transgenic chickens made to carry the CMV promoter driving the green fluorescent protein (GFP) exhibited transgene expression primarily in the pancreas and skin, and very weak expression in other tissues (McGrew et al. 2004). In addition, expression was not reported in some cell types, notably neurons, which exhibit strong transgene expression from the CMV promoter in mice (Van den Pol and Ghosh 1998). Another promoter used to drive constitutive transgene expression in mice, the promoter from the human ubiquitin-C gene, produces only low levels of transgene expression in quail (B.B. Scott and C. Lois, unpublished results). We have found two promoter/enhancer combinations that produce experimentally useful patterns of gene expression in transgenic quail. Using vectors carrying the human synapsin gene I promoter we have produced a line of transgenic quail with neuron-specific expression of GFP. Additionally, we found that the combination of the CMV enhancer and the chicken β-actin promoter produces ubiquitous expression in quail.

Before using a promoter/enhancer combination that has not been well characterized in birds, transcriptional activity should be tested either in cultured cells from that species, or by in ovo injection. Additionally, the transcriptional activity of a promoter may differ from one avian species to another. For instance, we have found that the CMV/β-actin promoter leads to high expression in quail, but very weak expression in zebra finches. Before the generation of transgenics in a novel species, candidate promoters should be tested to determine their activity in that species.

A second important consideration when choosing a promoter/enhancer combination is the possibility of side effects due to high levels of expression from ubiquitous promoters. Some constitutive promoters can overburden the normal transcriptional machinery of the cell, re-directing resources away from genes necessary for cell growth and towards the transgene. This slows cell proliferation, which puts infected cells at a selective disadvantage when compared to the naive, non-infected cells of the embryo, and lowers the germline transmission frequency of adult mosaics. We have observed this problem when generating transgenic quail that ubiquitously express GFP. When mosaic founder parent birds were generated with a lentiviral vector carrying the chicken β-actin promoter and CMV enhancer driving GFP, we found that only one out of ten founder quail produced transgenic offspring, and germline transmission frequency in those birds was low (<5% of progeny were transgenic). Somatic mosaicism in adult founders was also low (<1% of cells expressed GFP). In contrast, we generated transgenic quails in which GFP was driven by a neuron-specific promoter (human synapsin I), which showed very high levels of germline transmission. Germline transmission frequency in transgenic quail showing neuron-specific expression was between 5% and 33%, and five out of six mosaic quail produced transgenic offspring. Additionally somatic mosaicism in the neuron-specific transgenics was greater, (1–10% of neurons in these animals expressed GFP), than that of quail generated with the chicken β-actin/CMV vector (less than 1% of cells expressed GFP in the mosaic founders). Therefore, special care should be used when choosing promoters that may express in the germline of the developing bird, since the additional demands made on the cell's transcriptional machinery can impede transgenesis. Any manipulation that decreases the growth rate of infected germ cells, may ultimately lower the germline transmission rate of mosaic founders.

10.3.2 Transgenes

Up to 10 kb kb exogenous DNA, including cis-acting regulatory elements and protein-coding or other transcribed sequences, can be packaged between the flanking long terminal repeats of the lentiviral vector. Currently, only the protein markers GFP and beta-galactosidase, have been expressed in transgenic birds generated with lentiviral vectors. However, the range of possible transgenes is quite extensive, and limited only by the space constraints of the vector. Three general types of transgene are available for use with the lentiviral system. (1) Genes of interest can be ectopically expressed in the whole animal or selected tissues, based on the choice of promoter as described above. These might be endogenous genes, overexpressed to examine the effects of dosage on their normal protein function. Alternatively, exogenous genes that code for visible markers or biopharmaceuticals may be expressed. (2) Dominant negative constructs can be delivered that interfere with the function of endogenous genes (Herskowitz 1987). (3) Non-coding RNAs, in particular short interfering RNAs (siRNAs), can also be introduced using lentiviral vectors. siRNAs have been shown to be effective in silencing endogenous genes in transgenic mice generated using lentiviral vectors (Tiscornia et al. 2003). Importantly, the mouse U6 promoter, which is required to properly transcribe siRNA, has been shown to work effectively in birds (Katahira and Nakamura 2003; Pekarik et al. 2003). These strategies make lentiviruses useful vectors for genetic manipulation in birds.

10.4 Production of Transgenic Birds

10.4.1 Egg Preparation

Fertile eggs from many domestic species including quail, chicken, ducks and turkey are available by mail from farms or other breeders. Since bird embryos will start to develop when kept at room temperature, it is best to select a vendor that can guarantee freshly laid eggs and that can ship overnight. Vendors that specialize in avian research are preferable. We obtain quail eggs from CBT Farms (Chestertown, MD 2160). Orders are typically 100–200 eggs, which are shipped by express courier. If it is not convenient to inject eggs the day they arrive, eggs can be stored for up to 1 week at 15 °C in a cold room or refrigerator. When storing eggs for any length of time, we place eggs in a dedicated refrigerator and orient eggs so that the blunt end of the egg faces upward.

Once the eggs arrive in the laboratory, the crates are unpacked and the eggs placed in racks (Stromberg's Chicks, Pine River, MN; cat. no. ET124) and oriented so that the long axis of the egg is parallel to the ground. Keeping the eggs in this position allows the embryo to settle to the top of the egg so that it will be accessible in the window that will be created later for virus delivery. The eggs are left in racks at room temperature for 1 h, after which time we find that the majority of embryos lie just below the highest point of the shell. We mark this point with a felt tip pen to identify the presumptive location of the embryo.

For the windowing procedure, eggs are transferred to individual soft foam cradles. Using a hand rotary tool with a small bit, a circle with a 4 mm radius is drilled in the shell around the mark indicating the top of the egg. The window should be made shallow to avoid damaging the shell membrane. We have used both electric hand drills (Dremel) and air-powered dental drills (Benco Dental) to cut the window. Shell chips and dust can accumulate during the windowing. To eliminate this shell dust, dental drills can be purchased with a second channel that expels air at the head of the drill. This air stream is useful for blowing away the shell chips and dust that accumulate during drilling. If shell chips and dust remain after the cutting process, these should be wiped away with a Kimwipe moistened with 70% ethanol.

After the circular cut is made, the remaining shell fragments are removed with sterilized forceps to expose the shell membrane. Before removing the shell membrane, a drop of sterile phosphate buffered saline (PBS) large enough to cover the exposed shell membrane is added over the window. It has been shown that addition of PBS before removal of the shell membrane significantly increases hatchability (Speksnijder and Ivarie 2000). The shell membrane is then removed with forceps, taking care not to disrupt the vitelline membrane that surrounds the embryo and yolk.

10.4.2 Injection for the Production of Mosaic Founders (F0)

Once the shell membrane is removed, the embryo can be observed by the unaided eye. However, for precise targeting of the embryo we place the egg in a soft cradle under a dissecting microscope with 16X magnification.

Microinjection needles are fabricated from borosilicate glass capillary tubes (Sutter # B100-75-10) pulled using a Flaming/Brown micropipette puller, model P-97, using the following parameter settings:

$$\begin{aligned} \text{Heat} &= \text{Ramp} + 15 \\ \text{Pull} &= 30 \\ \text{Vel} &= 120 \\ \text{Time} &= 200 \\ \text{Pressure} &= 200 \end{aligned}$$

A ceramic tile (Sutter) is used to score and break the pulled glass give a blunt tip with a 20 μm outer diameter. Once cut, the microinjection needle is inserted into the pipette holder of a CellTram hydraulic injector (Eppendorf) and front-loaded with concentrated viral solution slowly (roughly 200 nl/s). To be able to visualize the viral solution as it is injected, we add the dye Phenol Red (final concentration 5%) before loading the viral solution into the pipette. We typically load 10–15 μl viral solution, mixed with Phenol Red, into a microinjection needle. A fully loaded needle should allow up to five egg injections.

High titer viral stocks are critical for the production of transgenic birds. We have successfully used viral stocks with titers of 10^6 to 10^7 infectious units per microliter (IU/μl) when titered on 293 cells. Using viral stocks of 10^5 IU/μl or lower will result in very low rates of germline transmission frequencies in mosaic birds. Previous studies using unconcentrated retroviral vectors (10^4 IU/μl) to produce transgenic birds have reported low germline transmission frequencies (<1%) in mosaic founders. In addition to using high titer vector stocks, it is important to minimize the time viruses spend in the injection needle, especially under halogen illumination, since vector particles degrade more rapidly at room temperature. To slow this degradation we keep the thawed viral aliquots on ice before loading into the microinjection needle.

Under a dissecting microscope, the microinjection needle is positioned above the center of the embryo using a micromanipulator. The needle tip is then advanced to penetrate the embryo, at an angle of approximately 45°. A Piezo Drill (Burleigh) attached to the microinjection needle can be used to help the needle tip rupture the vitelline membrane and penetrate the embryo. After the embryo has been penetrated, 3 μl vector solution is slowly injected into the subgerminal cavity, below the embryo. An injection is considered successful if the viral solution is observed spreading horizontally in a circle below the embryo and if the perimeter of the viral solution reaches the borders of the area opaca. In our hands more than 90% of injections are successful according to these criteria. Eggs with unsuccessful injections are discarded. For quails or chickens, we routinely inject 60–80 eggs per day. Assuming a hatching rate of 5–10%, this number of injected eggs will provide a sufficient number of founder animals for breeding.

Immediately after a successful injection, eggs are sealed in order to prevent microbial contamination and fluid loss during incubation. To seal the eggshell, a round glass cover slip (EMS, Ft. Washington, PA; cat. no. 72196–12) is placed over the shell window and is sealed to the egg with a biocompatible silicone elastomer (Kwik-Cast WPI). Eggs are placed blunt end up, in a forced air incubator (Brinsea) with a temperature of 38 °C and a relative humidity of 45% until hatching roughly 18 days later. The method of egg turning is an important factor to consider when choosing an incubator. Artificial incubator models that keep each egg (or eggs) in a single plastic cradle and change the orientation of the cradle, such as Brinsea and Alpha Genesis, are preferable to incubators that roll the eggs, such as Grumbach.

This simple method of sealing and incubating eggs typically yields a 10% hatch rate. Other incubation methods, both in artificial culture systems and in ovo, have been described that increase the hatching rate up to 30% or 60% (Perry 1988; Andacht et al. 2004). However, since germline transmission rates are high enough in mosaics produced with lentiviral vectors that only a few founders are needed to guarantee transgenic offspring.

10.4.3 Transgenic Offspring

Once mosaic founders have reached sexual maturity (7 weeks after hatching), these animals are bred to wild-type quail. Wildtype quail can be purchased from breeders and are usually inexpensive (<US $10 per adult quail). However, although the price for individual birds is typically low, shipping live birds can be quite expensive (US $300). Therefore, the best strategy is to order additional eggs and let them develop alongside the injected eggs, or alternatively, to breed mosaic founder quail to each other.

Since founder quail will be mosaic in all tissues, including the germ cells, only a percentage of their offspring will be fully transgenic. If the transgene produces a visible marker, such as GFP, the progeny can be screened for the presence of the transgene by phenotype. In general, however, screening is accomplished by either PCR or Southern blot analysis. Genomic DNA is required for both methods and can be easily extracted from the blood of birds. The alar vein of 5-day-old hatchlings is an excellent source of blood. The vein is nicked with a hypodermic needle and 70 μl blood is collected with a heparinized capillary tube (Chase Scientific Glass, Rockwood, TN). Genomic DNA can be extracted from blood by digesting overnight in proteinase K followed by phenol/chloroform extraction as described in (Sambrook and Russel 2001).

Lentiviral vectors integrate stably into the host chromosome and are passed through the germline to subsequent generations. Transgene expression is consistent from generation to generation; therefore positively identified transgenic birds can be used to establish breeding lines.

10.4.4 Husbandry

Husbandry concerns vary from species to species. Our laboratory has experience raising quails (*Coturnix coturnix*) and zebra finches (*Taeniopygia guttata*). Quails are the preferred species for testing injection methods, promoter/enhancers and general troubleshooting because they develop at a rapid rate (becoming sexually mature roughly 50 days after hatching) and because they are easy to house and care for. A colony of 30 breeding quails can be easily established by purchasing ready-to-assemble rack breeder cages (Stromberg's Chicks and Gamebirds Unlimited), and will only occupy 1.86 m^2 of laboratory space. In contrast, for most laboratories in an academic environment it will be difficult to establish a colony of chickens because their large size will require a special facility for their breeding. Zebra finches require more time to become sexually mature

(90 days) and must be raised by foster parents. Tutorials on quail husbandry with an agricultural or research focus are available (Randall 2001; BVAAWF/FRAME/RSPCA/UFAW 2001).

Our quails hatch in the incubator and are immediately transferred to a brooder (Brinsea) maintained at a temperature of 37 °C. After 5 days, the hatchlings are transferred to a larger battery cage that has been modified for hatchlings. The floors and sides have been reinforced with tighter mesh wiring (9 mm × 9 mm hole size) and the cage is equipped with an area heater (Start 'N Gro; Stromberg's Chicks and Gamebirds Unlimited) set to produce an average air temperate of 35 °C. Quail can also be raised in commercially available brooders, but our design can house both adults and hatchlings on different levels of a single rack of cages. Hatchlings are fed adult food that has been finely ground to a powder in a kitchen blender. Since quail hatchlings can drown in even shallow depths of standing water, we provided drinking water either in low plastic drinking fountains (Stromberg's Chicks cat. no. NDF) or in shallow tissue culture plates.

References

Andacht T, Hu W, Ivarie R (2004) Rapid and improved method for windowing eggs accessing the stage X chicken embryo. Mol Reprod Dev 69:31–34

Bosselman RA, Hsu RY, Boggs T, Hu S, Bruszewski J, Ou S, Kozar L, Martin F, Green C, Jacobsen F, Nicolson M, Schultz JA, Semon KM, Rishell W, Stewart RG (1989) Germline transmission of exogenous genes in the chicken. Science 243:533–535

BVAAWF/FRAME/RSPCA/UFAW (2001) Laboratory birds: refinements in husbandry and procedures. Fifth report of BVAAWF/FRAME/RSPCA/UFAW Joint Working Group on Refinement. Lab Anim 35[Suppl1]:1–163

Furth PA, Hennighausen L, Baker C, Beatty C, Woychick R (1991) The variability in activity of the universally expressed human cytomegalovirus immediate early gene 1 enhancer/promoter in transgenic mice. Nucleic Acids Res 19:6205–6208

Herskowitz I (1987) Functional inactivation of genes by dominant negative mutations. Nature 329:219–222

Katahira T, Nakamura H (2003) Gene silencing in chick embryos with a vector-based small interfering RNA system. Dev Growth Differ 45:361–367

Lois C, Hong EJ, Pease S, Brown EJ, Baltimore D (2002) Germline transmission and tissue-specific expression of transgenes delivered by lentiviral vectors. Science 295:868–872

McGrew MJ, Sherman A, Ellard FM, Lillico SG, Gilhooley HJ, Kingsman AJ, Mitrophanous KA, Sang H (2004) Efficient production of germline transgenic chickens using lentiviral vectors. EMBO Rep 5:728–733

Mozdziak PE, Petitte JN (2004) Status of transgenic chicken models for developmental biology. Dev Dyn 229:414–421

Pekarik V, Bourikas D, Miglino N, Joset P, Preiswerk S, Stoeckli ET (2003) Screening for gene function in chicken embryo using RNAi and electroporation. Nat Biotechnol 21:93–96

Perry MM (1988) A complete culture system for the chick embryo. Nature 331:70–72

Randall M (2001) Raising Japanese quail PhD thesis State of New South Wales, Department of Primary Industries, Agriculture

Sambrook J, Russel DW (2001) Molecular cloning: a laboratory manual. Cold Spring Harbor Laboratory Press, Cold Spring Harbor, NY

Schmidt EV, Christoph G, Zeller R, Leder P (1990) The cytomegalovirus enhancer: a pan-active control element in transgenic mice. Mol Cell Biol 10:4406–4411

Speksnijder G, Ivarie R (2000) A modified method of shell windowing for producing somatic or germline chimeras in fertilized chicken eggs. Poult Sci 79:1430–1433

Tiscornia G, Singer O, Ikawa M, Verma IM (2003) A general method for gene knockdown in mice by using lentiviral vectors expressing small interfering RNA. Proc Natl Acad Sci USA 100:1844–1848

Van den Pol AN, Ghosh PK (1998) Selective neuronal expression of green fluorescent protein with cytomegalovirus promoter reveals entire neuronal arbor in transgenic mice. J Neurosci 18:10640–10651

11 Ancillary Techniques

Shirley Pease

11.1 Introduction

The information provided in this chapter relates to the use of rodents in the generation of genetically altered animals. All the topics written about here are well documented in other texts. The purpose of this chapter, then, is to provide the user with the fundamentals on each topic, such that this manual stands alone as a research tool. References are given to enable the user to further research each topic. Another recommended resource would be an email list at e-mail: transgenic-list@imperial.ac.uk (owner, Peter Sobieszczuk). This is a website for personal exchange of information and experiences, provision of advice, and troubleshooting on all topics related to and covered in this chapter.

11.2 Getting Started

One person can sit at a microscope and inject embryos on 3 days of each week. More than that becomes taxing in many ways. The experienced injector can inject 100 blastocysts, or 300 single cell embryos (zygotes), in 1 day. These are enough embryos for implantation into six to nine recipient females each day, the number being determined by embryo survival rate, particularly where pronuclear injected single-cell embryos are concerned. The number of embryos that need to be injected may also be influenced by your choice of methodology in making genetically altered animals. To generate a stable transgenic line, selecting from more than one founder animal generated by pronuclear injection of zygotes, may require 3 days of microinjection. To make a transgenic animal by lentiviral injection will take only 1 day of microinjection, although establishing a stable transgenic strain may require more breeding steps later on. In this chapter, I will de-

Shirley Pease: California Institute of Technology, Division of Biology, Pasadena, CA 91125, USA, e-mail: ssp@its.caltech.edu

Principles and Practice
Mammalian and Avian Transgenesis – New Approaches
Shirley Pease and Carlos Lois (Eds.)

scribe the colonies of mice and rats and the equipment needed in order to complement the work of one microinjectionist.

11.2.1 Microinjection Equipment

A dissecting microscope, such as a Nikon SMZ-800 on a boom stand with ring light and zoom magnification, for use in the vasectomy of male mice and/or the implantation of embryos into recipient females. Include 10X eyepieces with a zoom to 40X.

A second dissecting microscope, such as the Wild M3C, with a transmitted light base, 16X eyepieces and 0.64X, 1X, 1.6X and 4.0X objectives. The transmitted light base allows you to fully examine the embryos for quality. Alternatively, a Nikon SMZ-1500 (zoom lens) with a heated base (an SMZ Thermo Plate with Controller from Biogenics, Napa, CA; www.biogenics.com).

A pipette puller, such as model P-97 from Sutter Instruments (Sutter Instruments, Novato, CA; www.sutter.com) for the production of injection and embryo-holding pipettes. Alternatively, pre-fabricated pipettes may be purchased (Eppendorf, Hamburg, Germany; www.eppendorf.com; Humagen, Charlottsville, VA; www.humagenivf.com; Biomedical Instruments, Zoellnitz, Germany; www.bio-medical.com).

A microforge such as that made by Narishige (Tokyo, Japan; www.narishige.co.jp/main.htm), for the shaping and finishing of microinjection and holding pipettes, pulled on the Sutter.

An inverted microscope, such as those made by Zeiss, Leica or Nikon, for the injection of embryos. You may need both phase and Nomarski or Hoffman optics. Nomarski or Hoffman modulation optics are essential for pronuclear injection. Blastocyst injections can be accomplished with a simple phase contrast microscope, as can sub-zona injection of lentivirus. Select 10X, 20X, and 32X or 40X objectives (32X is good for blastocyst injection and sub-zona injection; 40X is most suitable for pronuclear injection).

One anti-vibration table to go under the microinjection scope. Vibration from other building equipment when working at such high magnification is a common problem, especially if the room in which the scope is located has floors below. It is possible to reduce the effect of vibration by sitting the scope on a table built from substantial marble slabs, the tabletop supported in its position across the uprights by layers of rubber cushioning. A superior solution can be found in an air-cushioned isolation table, such as made by Technical Manufacturing Corporation (Peabody, MA;

www.techmfg.com). You will need a high-pressure in-house air supply line or a cylinder supply with such tables.

One set of micromanipulators, for holding and manipulation of microinjection and embryo-holding pipettes. Eppendorf motorized models are very nice, and suitable for all types of injection, but are comparatively expensive. Narishige remote joystick manipulators work well and are suitable for all types of injection. Leitz manual micromanipulators are more suitable for the injection of blastocysts and for sub-zona injections than for pronuclear injections, although many researchers use them successfully for the latter purpose.

Hydraulic micrometers, each system needs two, for controlling the movement of embryos and either cells or viral preparations on the microinjection set-up. The Eppendorf CellTram Oil or Vario are very nice, and are essential for lentiviral injection. The Narishige is a less expensive alternative, well suited to embryonic stem (ES) cell injection and all embryo-holding requirements. An air-filled micrometer syringe from Zandar Medical Supplies (www.zandar.com) is another alternative for embryo holding and ES cell injection.

For DNA injection, we recommend a compressed air system such as Femtojet, by Eppendorf, or the Picospritzer (General Valve Corporation, Fairfield, NJ). The advantage of the Femtojet is that it has a built-in pump, which provides compressed air. The Picospritzer is less expensive, but requires connection to an air supply line for operation.

11.2.2 Mouse Stocks for Embryo Production and Implantation

11.2.2.1 Hormone-Primed Embryo Donors

Stocking requirements: 45 stud males for repeated use and 30—45 donor females for each group of 3 injection days.

Most commonly used for lentiviral and pronuclear injection are hybrid animals such as B6D2F1, B6CBAF1 or B6C3F1. Also used are inbred strains such as FVB/N mice or C57BL/6. The latter may be selected for use because of specific features of their genetic background. Hybrid animals are used because, due to hybrid vigor, they produce larger numbers of embryos in response to hormone priming, and pup yields from injected hybrid embryos tend to be higher. You will need a permanent stock of stud males. These animals should be singly housed for repeated use. If males are used too frequently, their level of fertility may temporarily fall. Ideally, use each male no more than once every 3 days. Thus, between 30 and 45 single stud males should be enough. If using these males to mate with hormone-primed females, then 10–15 hormone primed females, mated 1:1 with stud males, should provide enough embryos for manipulation. Therefore, if embryos are required for injection on 3 consecutive days of each week, it will be necessary to use each male only once a week. One hormone-primed female will produce between 3 and 30 embryos suitable for injection (see Sect. 11.3.2).

11.2.2.2
Naturally Mated Embryo Donors

Stocking requirements: 90 stud males for repeated use and 100 or more donor females, for each group of 3 consecutive injection days.

It is possible to obtain embryos from non-hormone-primed females. In fact, for the collection of blastocysts, some prefer to use non-hormone-primed females, since the quality of the blastocysts collected at day 3.5 post-coitus (p.c.) is sometimes superior to those obtained by hormone priming, although their number will be lower. If using naturally mated females, which will produce between three and ten embryos suitable for injection per female, then mate females with studs at a ratio of 2:1. Due to the 4–5 day oestrus cycle in female mice, calculate that only one-fifth of paired females will be in oestrus, and will therefore mate, each night.

It is possible to increase the percentage of females that will mate, to some degree, by examining females for physical signs of oestrus before setting up for mating. The vulva of the female will appear reddened and swollen during oestrus (Rough 1993). Even under these circumstances, in our experience, only a maximum of 50% of females are likely to mate. Thus, if you naturally mate 60 randomly chosen females (or 30 females showing signs of oestrus), you can expect to recover about 40 blastocysts from an inbred strain or 80 single cell embryos from a hybrid strain. Remove the females the following morning, checking for plugs as you go.

11.2.2.3
Recipient females

Stocking requirements: 60 sterile males (either vasectomized or sterile because of a gene mutation) for repeated use, and 60 females available for mating each week.

For the production of pseudopregnant recipient mice, you will need a stock of sterile males and fertile females. Even though fertile embryos will not be produced in these matings, the physical stimulation of mating brings about hormonal changes in the female that would naturally occur after a fertile mating. This is called pseudopregnancy and, in a female that is not used as a recipient animal, may last for 14 days. After this time, the female will start to cycle normally again. However, during pseudopreg-

Table 11.1. Pups Born after Implantation into Naturally Mated Females

Number of embryos implanted	Embryo stage	Total number pups born	Number from host female	Number from implanted embryos	Percentage of litter
9	Blastocyst	8	7	1	12.5%
19	Blastocyst	12	9	3	25%
19	Blastocyst	8	5	3	37.5%
19	Zygote	10	6	4	40%
19	Zygote	10	5	5	50%
17	Zygote	4	3	1	25%
37	Zygote	15	6	9	60%
30	Two cell	9	4	5	55.5%
23	Eight cell	9	5	4	44%
Total 192		Total 85	Total 50	Total 35	Av. 41%

nancy, the female will not go into the oestrus phase and therefore will not mate. This is an important point to remember where plugged but unused females are returned to the mating pool for re-use. The strain of mouse chosen for such a purpose is important. A fertile strain with good body size and behavioral characteristics is essential, such that when recipients give birth to manipulated embryos, the recipients are well able to rear large numbers of pups. We highly recommend the use of outbred (CD-1, ICR, Swiss Webster) or hybrid (e.g., B6C3F1 or B6CBF1) strains as foster mothers, due to their ability to produce and raise larger litters more effectively than inbred strains. The strain of the vasectomized or sterile male does not really matter, as long as the animals are able to mate efficiently.

It is also possible to use females mated with fertile males as recipients. However, this is not as efficient as the use of pseudopregnant females, since litters born will consist of both pups arising from the implant of injected embryos and pups arising from the natural mating of the recipient female. Thus the total number of injected embryos that give rise to pups will be less. Nevertheless, there may be circumstances under which the use of fertile males may be used (see Sect. 11.4). Under these circumstances, expect to see that about 41% of a litter born is from transferred embryos (see Table 11.1).

After the injection of 200 zygotes or 60 blastocyst stage embryos per day, and depending upon the stage at which the embryos are returned to the uterus, you may need up to six pseudopregnant recipient females per day of injection. In this case, plan to have eight available, since not every female apparently mated appears perfectly timed for the implantation of embryos, according to the appearance of the uterus, oviduct and ovaries at the time of surgical implantation. It is a waste of embryos and your work to implant embryos into recipients that do not appear to be correctly timed. So try to produce more recipients than you need, to allow yourself the opportunity to use only the best.

In order to produce eight pseudopregnant females, and given the 4–5 day oestrus cycle, five times the number of females will need to be mated, at a ratio of two females to one vasectomized male. Thus you will need 20 sterile males for use on each of three nights per week. Although fertility is not an issue here, the males can become tired or disinterested if overused, resulting in lower mating efficiency. Therefore, continue the practice of using any male no more than once every 3 days. Hence, a group of 60 sterile males at 2–9 months of age should be maintained. Sterile males can be obtained in any of three ways. You can purchase males that, due to genetic background, are sterile from birth, such as those from the T145H-Re strain of mice from Harlan Sprague Dawley (Indianapolis, IN; www.Harlan.com). This strain is reported to be comparable to a hybrid mouse for mating activity (Robinson et al. 2003). Alternatively, you can vasectomize males yourself, by surgically removing a small piece of each of two vas deferens from each male, as described by (Nagy et al. 2003). It is also possible to purchase mice that have been surgically altered in this way from a commercial supplier of animals, such as Charles River Laboratories (Wilmington, MA; www.criver.com).

As mentioned above, you will need to mate 40 randomly selected females each night, in order to obtain an average of 8 recipient females each morning. So plan to keep a stock of up to 60 females, between 2 and 5 months of age, available for use at all times. It is possible to reduce the number of females to be mated by selecting only those that show external signs of oestrus (Champlin et al. 1973; Rough 1993). In this case we would recommend mating 16 such chosen females in order to produce 8 timed mated females the following morning. If making a practice of selecting for mating only females in oestrus, it would be possible to maintain fewer vasectomized or sterile males.

11.2.3 Rat Stocks for Embryo Production and Implantation

11.2.3.1 Embryo Donors

Stocking requirements for lentiviral injection: six stud males for repeated use (no more than once every 3 days), and six females per day's injection.

We have superovulated rats for the production of embryos for lentiviral injection. The introduction of foreign DNA into the genome by this method is much more efficient than the introduction of DNA by pronuclear injection (Lois et al. 2002). Thus the superovulation of small numbers of female rats has been sufficient to produce enough embryos for the work. In this case, it is necessary to hormone prime only six suitable females, each of which will produce 20–30 embryos suitable for injection.

11.2.3.2 Recipient Females

Stocking requirements: either 20 vasectomized male rats and 40 stock females, all between 12 and 20 weeks of age, or 6 vasectomized males and 6 timed mated females.

The logistics of producing pseudopregnant recipient female rats is the same as that of mice, their oestrus cycles being of very similar duration. Thus, a large number of sterile males could well be needed. However, this may not be very practical, because the housing of rats requires much more space, and is thus more costly, than the housing of mice. The number of sterile males needed could be reduced, as for mice, by selecting for mating only those females that show external signs of being in oestrus or by pre-screening candidate females by vaginal smear. (The criteria for establishing oestrus in the female rat are identical to those for mice, see (Rough 1993).) It is also possible to synchronize oestrus in a group of females by giving human chorionic gonadotrophin (hCG) 4 days prior to the evening of mating to vasectomized males (Mullins et al. 2003). Another approach would be to purchase recipient females that have been timed mated to litter 1 or 2 days before the day of microinjection. Specifically, if injections are due to be done on a Friday, purchase females due to litter on Wednesday and Thursday of the same week. Post-partum oestrus will provide the opportunity for mating within 24 h after littering. Post partum oestrus begins 12–20 h after parturition and lasts for 12 h. On Thursday evening, post parturition females should each be mated with a vasectomized male, at a ratio of 1:1. It is possible also to pair up any female that you believe will litter during the coming night.

11.3 Embryo Production in Rats and Mice

The use of many of the techniques described in this manual requires the production of larger than normal numbers of embryos from each donor female. This section outlines the parameters for obtaining such embryos from various strains of mouse and rat. At the time of publication, only a small amount is known about varying responses to hormone priming in different strains of rat.

Each strain of mouse or rat may respond differently to hormone priming. One aspect of superovulation that is common to all rodents is that the animals must be held under controlled light systems. Fluctuating or suboptimal exposure times to light and dark cycles may exert an influence over the oestrus cycles of rodents, confounding the effects of the administration of exogenous hormones. Therefore, for optimal animal management, it is critical to keep the light cycles constant, so that the animals will not respond to seasonal variations in light duration and intensity, and thus will continue to be able to reproduce all year round (Guide for the Care and Use of laboratory animals, 1996). Rodents may be successfully maintained under conditions that provide 12 h light and 12 h darkness in a 24 h period. Alternatively, they may be maintained under 14 h light and 10 h darkness. The latter is most commonly used. Currently, our own animal facility lights are set to turn on at 6:00 a.m. and to go out at 8:00 p.m. each day. Thus, the mid-point of the dark cycle is 1:00 a.m. This is a critical point, since ovulation and endogenous hormone levels are determined by the staging of the period of darkness, not the period of light.

Therefore, in order to exert an influence over hormone levels and subsequent ovulation, it is important to deliver exogenous hormones at a time during the 24 h period when they will be effective, and this will be in relation to the mid-point of the dark cycle. Endogenous hormone levels may confound the effect of exogenous doses. Thus it is important to deliver exogenous doses at optimum times. For example, pregnant mare serum gonadotrophin (PMSG) should be given at a time prior to that when endogenous levels of the hormone are rising in the bloodstream. Once endogenous hormones have had an effect, it is difficult to enhance their action with exogenous doses given after the event. Details of hormone priming regimes that have worked for others are given in Tables 11.2 and 11.3. (It should be noted that, although the variations in dose, time and age of donor varies, sometimes by small amounts, there is no statistical data to support the theory that these minor variations produce real effects in embryo yield – our intention is to provide the reader with a range of starting points with which to work.)

PMSG contains large amounts of follicle stimulating hormone (FSH). This promotes the maturation of large numbers of ova from the ovary. hCG contains high levels of luteinizing hormone (LH), which stimulates the release of mature ova from the ovary. In general, hCG should always be given 46–48 h after the delivery of PMSG. The timing of the PMSG dose should be such that it will enhance endogenous hormone levels, the rise and fall of which relate to the mid-point of the dark cycle. When planning such work, keep in mind also that mature females, if purchased from a supplier, may be coming from a different light cycle, or even a different time zone. Thus they will need a minimum of 3 days to acclimate to your in-house light cycle, and to recover from the stress of shipping, where this is applicable.

Table 11.2. Hormone priming of mouse strains. *PMSG* Pregnant mare serum gonadotrophin, *hCG* human chorionic gonadotrophin

Strain	Age and/or weight	PMSG dose (IU)	Timing – hours before mid-point of dark cycle	hCG dose (IU)	Timing – hours after PMSG dose
BALB/c	5 weeks	5	10	5	45
BALB/c	6–8 weeks	5	10	5	47
BALB/c	9–10 weeks	5	12	5	48
B6D2F1	8 weeks	5	12	5	46
CD-1	10 to 12 g	5	9	5	46
CD-1	3 to 4 weeks	7	12	5	47–48
CD-1	4 weeks	5	7–11	5	46–48
CD-1	5–6 weeks	5	10	5	47
C57BL/6	21–24 days and 9–11 g @ 21 days	5	12	5	46
C57BL/6	21–25 days	5		2.5	
C57BL/6	25–28 days	2.5	12–13.5	2.5	47–48
C57BL/6	28–42 days	5		5	
C57BL/6	5 weeks	10	14	10	46
FVB	4 weeks	5	11.5	7.5	46
FVB	5 weeks	5	8.5	5	44–48
FVB	7–9 weeks	5	7–9	5	46–48
NOD	21 days	4.5	13	3.5	46
NOD/LtJ	11–12 weeks	8.75–10	8.5	8.75–10	46.5

Table 11.3. Hormone priming of rat strains (hormones administered by intraperitoneal injection)

Strain	Age and/or weight	PMSG dose (IU)	Timing – hours before mid-point of dark cycle	hCG dose (IU)	Timing – hours after PMSG dose
F344	28 days, 70–80 g	15	7–9	5	48
F344		15–40		5	
F344	5 weeks	40	13	30	49
F344	28–30 days and 70–80 g	25	13	5	48
Sprague Dawley	30–35 days, 80–90 g	30		25	52
Sprague Dawley	30–35 days	30		25	52
Wistar		20		7.5	46
Wistar	4–5 weeks	20		20	50–52
Wistar	10–15 weeks	150		75	48
CD-1	8–10 weeks	10	12	10	47
NOD/LtJ	8–9 weeks	5	9	5	46

11.3.1 Preparation of Hormones and Enzymes

Hormone preparations can be obtained from several sources. Options include: Calbiochem (La Jolla, CA; www.bioresearchonline.com), Sigma (St Louis, MO; www.sigmaaldrich.com), National Hormone and Peptide Program (Torrance, CA; www.humc.edu/hormones), and Professional Veterinary Products (Omaha, NE.; www.pvpl.com).

Hormones should be aliquoted at 25 IU (international units) per milliliter in phosphate-buffered saline (PBS) for use in mice and rats. Aliquot in quantities tailored to your daily use requirements. Hormone preparations should be stored at −20 °C. They will be stable for 2–3 months at this temperature. Once thawed, working stocks should be immediately used and/or discarded. PMSG is sensitive to freeze/thaw cycles and should not be stored in a frost-free freezer.

Hyaluronidase is an enzyme naturally produced by sperm to assist in the penetration of the cumulus mass cells surrounding ova after their release from the ovary. A pig testes preparation of the enzyme is commercially available from Sigma (catalogue number H3884). This preparation is presented as 30 mg of a lyophilized form of the enzyme. This should be dissolved in 100 ml M2 medium, for a final concentration of 300 μg/ml, then stored in approximately 4 ml aliquots at −20 °C until needed for use. This enzyme will be stable for several years at this temperature.

11.3.2 Hormone Priming of Mice

For the collection of one-cell embryos from mice, PMSG should be given on day -2.5, day +0.5 being the day of collection, as shown in the work flow plan below. Typically, 5 IU PMSG should be given intraperitoneally (IP) at a period between 1:00 p.m. and 3:00 p.m. on day -2.5 and 5 IU hCG by the same route, 46 h later on day -0.5, where the mid-point of the dark cycle is at 1:00 a.m (see Table 11.2 for hormone priming regimes for mice).

Immediately following dosing with hCG, or before the onset of the dark cycle, hormone primed females should be paired with males at a ratio of 1:1. One-cell embryos from females that have mated may then be collected the following morning. Blastocyst stage embryos are collected 3 days later.

Where one-cell embryos are required for injection, the numbers needed may vary. For pronuclear injection in the generation of transgenic mice, 200–300 embryos may be needed per day of injection. In this case, 10–15 females should be hormone primed and mated. Generally, hybrid females will produce more eggs per donor than females of inbred strains. It is not practical to try to approach this type of microinjection procedure without hormone priming donor females, since relatively large numbers of embryos are needed. For lentiviral work, fewer embryos are required, perhaps only 100. In this case, scale down the number of donors to five or six mice.

Workflow plan (where day of microinjection = day 0.5):

	Zygote production	Blastocyst production
Day −5.5		Give PMSG
Day −3.5		Give hCG and mate females
Day −2.5	Give PMSG	Plug-check females and separate from males
Day −0.5	Give hCG and mate females	
Day +0.5	Plug-check females. Harvest zygotes from ampullae of plugged females	Harvest blastocysts from uteri of plugged females

Where blastocysts are needed, hormones should be given such that the animals are mated 4 days before blastocyst injections are to be done, blastocysts being harvested on day 3.5, the morning after mating being day 0.5 (see work flow plan). Ten hormone-primed females of optimum age and weight for the strain (see Table 11.2) should provide enough blastocysts for a good day's work. It is possible to harvest enough blastocysts for injection from non-hormone primed females (see Sect. 11.2.2). In this case, if working with C57BL/6 mice, calculate to harvest only three good blastocysts from each plugged female, and calculate that no more than 20% of paired females will mate. Thus, if you need 40 blastocysts to work with, plan to naturally mate 70 females. Figure 11.1 is a useful organizational table/work request form that you may wish to use when setting up the matings of donor and recipient rats and mice.

11.3.3 Hormone Priming of Rats

For this species, natural mating for embryo production seems impractical, given the number of rats that would need to be set up for mating. Nevertheless, if this is an option, apply the criteria as for mice, since their oestrus cycles are very similar. In hormone priming, outbred rats have been seen to give a better response in the number of embryos produced than hybrid or inbred rats (Corbin and McCabe 2002). In hormone priming of rats, it is possible to use a small pump, implanted subcutaneously, which will deliver

TIMED MATING REQUEST FORM REF. #

Hormone-primed females (1 male : 1 female)																
Females				0.2 ml (5iu) PMSG (1 to 3 PM			(46-48 hrs later) 0.2 ml (5iu) HCG			(After HCG, mate inmediately with) Male						
Room	How many	Strain	Age	Date	Time done	Init	Date	Time done'	Init	room'	Strain	Date (by 10 AM) of plug check	Init	Number of plugs obtained	Date to send to Lab	Requester

Naturally-mated females (1 male : 2 females)											% of plugs	% of plugs
	Strain								Number of plugs			
Room	Females	Males	How many females to mate	Date to mate	Init	Date(by 10 AM) of plug check	Init		needed	obtained	Date to send to Lab	User/ purpose (PI and/ or tech)

Production of pseudopregnant recipient females:- mate females with vasectomized males (1 male : 2 females)										
Room	Strain of females	How many females to mate	Date to mate with vas male	Init	Date (by 10 AM) of plug check	Init	needed	obtained	Date to send to Lab	User/ purpose

Tech. Request

Fig. 11.1. Timed mating request form

exogenous hormone over an extended period of time. It is also possible to deliver the hormone by IP injection and still achieve good results. Many factors will play a part in this, such as the strain of animal and the age and bodyweight at which the hormone is given to each donor animal. See Table 11.3 for hormone priming regimes for rats, administered by injection only.

11.3.4 Embryo Collection and Culture in Mice

Immediately following dosing with hCG, or before the onset of the dark cycle, hormone-primed females, mice or rats, should be paired with males at a ratio of 1:1, or non-hormone-primed females mated at a ratio of two females to each male. The following morning, each female should be examined for the presence of a copulatory plug, formed by part of the ejaculate from the male. The presence of a plug indicates mating has occurred. Females should be checked for plugs before 11:00 a.m. in the morning, since as time goes by, the plug will dry, shrink, and may fall out. Separate females from males at this time. In describing the developmental stage of embryos, we start from the premise that mating occurred at the mid-point of the dark cycle. Thus, the morning of plug check is day 0 or embryonic (E) day 0.5, or p.c. day 0.5. Zygotes should be collected from mice on this day. Blastocysts can be collected on day E3.5.

The timing of the collection of zygotes is important. It is easier to collect one-cell embryos while they are surrounded by cumulus mass cells and still in the ampulla of the oviduct. If collection is left too late in the day, then it is likely that the cumulus mass cells will start to fall away from the embryos, and the embryos will begin to move down the oviduct, making it more difficult to isolate them from the oviduct. When injecting into the pronucleus of embryos, it is most practical to do so when the pronuclei are at their largest. In lentiviral injection, this is less critical. In either case, it is important to inject the zygotes before the cell is about to divide, or the generation of animals mosaic for the transgene is more likely. Plan to isolate one-cell mouse embryos between 8:00 a.m. and 11:30 a.m.

11.3.4.1 Tools and Equipment

- Scissors, 11 cm, fine pointed straight; 1 pair
- Forceps, 11 cm, serrated, straight; 1 pair
- Forceps, 11 cm, fine pointed watchmaker's number 5; 2 pairs

- 37 °C warm block
- Dissecting microscope with transmitted light base
- Mouth pipette and glass embryo transfer pipettes (see Appendix 1), or Drummond Dialamatic Sequencing pipette (Drummond Scientific, Broomall, PA; www.drummondsci.com, Cat. No. 3-000-203)
- Light liquid paraffin oil (filtered to 0.45 microns and batch tested for toxicity to embryos)
- Media – M2 or equivalent (HEPES-buffered for bench work)
  - KSOM or equivalent, (bicarbonate buffered, for embryo culture in CO_2 incubator; both available from Specialty Media, Phillipburg, NJ 08865; www.specialtymedia.com)
- Enzyme – hyaluronidase, bovine type IV-S (Sigma, cat no. H 3884). Use at 300 μg/ml in M2, stored at −20 °C, indefinitely.
- Plastic ware – bacterial Petri dishes, such as made by Falcon, 35 mm.
- Plastic transfer pipettes. (Samco Scientific, San Fernando, CA 91340, cat. no. 202-15)

11.3.4.2 Collection of One-Cell Embryos

To visualize the embryos, once collected from donor animals they must be treated with hyaluronidase to free them from the cumulus mass cells that surround them for some hours after their release from the ovary. During this procedure it is important to work with the embryos at 37 °C wherever possible. Therefore, minimize the time the embryos are out of the incubator and warm media before use. Work on a warmed microscope stage where possible.

1. Warm one aliquot each of M2 and KSOM media and hyaluronidase on the 37 °C warm block. Prepare a 35 mm Petri dish with a small volume (1–2 ml) of warmed M2 medium.
2. Sacrifice embryo donors, either by CO_2 inhalation or cervical fracture (Howell et al. 2003).
3. Flood the ventral surface of each donor's body with 70% ethanol.
4. Make a central incision, first through the skin and then through the muscle of the abdominal wall. Expose one uterine horn. Pick up the ovarian fat pad and put tension on the uterine horn. This will separate

the ovary from the oviduct, although each will still be held in close proximity to the other by the bursa.

5. Pick up the uterine horn at the end near the oviduct. Cut the uterine horn just below the forceps and then between the ovary and oviduct. Put the oviduct into the Petri dish, in M2.
6. Repeat, collecting all available oviducts in one Petri dish.
7. Under the dissecting scope, take one oviduct at a time and place in a clean drop (500–1,000 microl) of M2. With two pairs of fine watchmaker's forceps, locate the ampulla and tear it apart. In most cases, clumps of cumulus mass cells surrounding and containing the zygotes will be seen to emerge from the ampulla. Sometimes, single zygotes can be found. Work through all available ampullae, freeing all cumulus masses into one drop of M2.
8. Either by mouth pipette or with the Drummond sequencing pipette, collect up all the cumulus masses and free zygotes you can find. Place them in a dish containing warmed hyaluronidase (~ 2 ml). The cumulus mass cells may be seen to fall away from the embryos over the next 5 min. It may be necessary to pick up all the embryos once or twice and move them to a new position in the dish of hyaluronidase, in order to free them from all the cumulus mass cells.
9. As soon as the zygotes appear to be free of cumulus cells, transfer them into clean medium. If you are going to start working with the embryos, then wash them through several drops of M2 medium, to remove all the hyaluronidase from their environment. If you need to return them to the CO_2 incubator, then wash them through several drops of KSOM. Culture the embryos in a small drop (50 μl) of KSOM, covered with oil, in a 35 mm Petri dish. In this case it is important to free the embryos from all traces of hyaluronidase and M2. Overexposure to hyaluronidase may harmful to the embryos, since this enzyme is a mitogen. Overexposure may also make the embryo zonas sticky and thus difficult to work with. Traces of M2 in the incubator culture will prevent the embryos from developing at all, because if medium containing HEPES buffer is exposed to CO_2, the pH of the medium will become more acidic. Single cell embryos are very sensitive to pH and temperature. If either is not exactly right, then the embryos will fail to develop to two-cell stage.

11.3.4.3
Collection of Blastocysts

Hyaluronidase is not needed in this case. Work through steps 1–5 above.

6. Pick up uterine horn at the end near the cervix and cut across the horn, close to the cervix. Place horn, with oviduct, in M2 medium. Collect all uterine horns.
7. Under the dissecting scope, take one horn at a time and flush 500 μl medium through each horn using a 1 ml syringe fitted with a 25 g needle. Hold the horn with fine forceps at the oviduct end. Insert the flushing needle into the lumen of the horn, also at the oviduct end. At this stage of development, the embryos should have just moved into the lumen of the uterus from the oviduct and thus should be located in this area.
8. Flush all horns into one or more dishes. Finally collect all embryos by mouth pipette or Drummond sequencing pipette, wash well in KSOM and place in microdrop culture of KSOM under oil.

11.3.5 Embryo Collection and Culture in Rats

One-cell rat embryos should be harvested after midday, since their developmental ability is known to be very low when harvested for use soon after penetration of the egg by the sperm (Miyoshi et al. 1997). Plan to harvest these embryos between 12:00 midday and 1:00 p.m. These embryos can be successfully cultured for short periods of time in media developed for the culture of murine embryos. In rats, osmolarity and adequate levels of available glucose in the medium seem to be critical factors in successful culture from two-cell stage through to blastocyst stage. A medium for this purpose, modified rat 1-cell embryo culture medium (mR1ECM), which will allow 80–90% of embryos to go to blastocyst stage, has been developed (Miyoshi et al. 1995, 1997).

11.3.5.1 Tools and Equipment

Assemble tools as in Sect. 11.3.4.1. In addition:

- Scissors, 12 cm, straight pointed, 1 pair
- Serrated forceps, 12 cm, straight, pointed, 1 pair

11.3.5.2 Procedure

The embryo collection procedure in rats is very similar to that of mice. Follow steps 1–10 as for the collection of one-cell embryos from mice.

Anatomically, there are a few differences. The oviduct and infundibulum is much more closely connected by the bursa to the ovary in rats than in mice. In mice, it is possible to separate the oviduct from the ovary quite clearly, by moving the uterine horn away from the ovary. In rats, it is not possible to see the separation so clearly. The ovaries of some strains of rat are disproportionately larger than those of mice.

11.4 Transfer of Mouse and Rat Embryos

In preparing mice as recipients of manipulated embryos, it is usual to use female mice that are pseudopregnant. These are females that have undergone mating with a sterile male (see Sect. 11.2.2). As mentioned in Sect. 11.2.2, it is also possible to use females that have been mated to fertile males as recipients for manipulated embryos. In this case, a mixed litter will likely be born, consisting partly of pups developed from injected embryos and pups arising from the natural mating, a disadvantage in that smaller numbers of pups developing from manipulated embryos are likely to be born (see Table 11.1). In addition, if these pups are genetically disadvantaged in some way, due to the effects of the transgene, then they may be further challenged by having to compete with healthy, non-transgenic littermates. However, for small or logistically difficult projects, such as for implanting embryos into axenic (germ-free) recipients, where the process of surgically vasectomizing the males may risk compromising the microbiological status of the animals, the possibility of working with un-vasectomized males may be of significant advantage. It should be kept in mind, though, that the number of pups born from the transfer of embryos will be reduced, per implant. Our experience has been that, in litters where fertile heterozygote CD-Nu males and females have been used to generate suitable recipient females and embryos transferred have been from immune deficient strains, approximately 41% of pups born have developed from implanted embryos.

11.4.1 Synchronization and Implantation of Recipient Mice

There are two routes for the transfer of embryos into recipient mice. They may be transferred into the oviduct via the ampulla or the infundibulum, or into the uterine horns of such animals. In normal development, the embryo first makes contact with the uterine wall at E3.5, and implantation takes place at E4.5 to E5.0. It is difficult to implant into the middle of the oviduct, because here the oviduct walls are less manipulable than in the area of the ampulla, and there is also less available space in the lumen of the oviduct. However, it can be done (see Chap. 4, *Cloning the Laboratory Mouse by Nuclear Transfer* by K. Eggan and R. Jaenisch). It is important to synchronize the developmental stage of the embryos with the intrauterine conditions of the recipient mouse. Thus, all stages of embryo from one-cell to pre-compacted morula stage (E0.5–E2.5) may be transferred into the oviduct of day 0.5 p.c. recipients. Compacted morulae and blastocysts may be transferred either into the uterus of day 2.5 p.c. recipients, or into the oviducts of day 0.5 p.c. recipients. Recipient mice that were mated the previous night and found to have a plug the following morning are suitable for oviduct transfers on the day of plug. However, we have found that the more disparate the embryos and recipients become, the less efficient the procedure (S. Pease, unpublished data). Others report that they regularly use day 0.5 p.c. recipients for all embryo transfers, from E0.5 to E3.5 stages, with no reduction in efficiency. The asynchrony of 24 h (or more) allows implanted embryos time to recover from the manipulation procedure and/or sub-optimal embryo culture conditions, either of which may delay the development of the embryos. Embryo development will continue once the embryos have been returned to the uterus and the resulting blastocysts will wait until the uterine wall of the recipient, technically 24 h behind the developmental stage of the embryos, is ready for the next stage: implantation of embryos. If E3.5 embryos were returned to the uteri of day 3.5 p.c. recipient mice, it is likely no pregnancy would result. This is because the embryos, delayed in development because of their manipulation and in vitro hours, would not be ready for implantation at the same time as the uterus. The window of opportunity would be missed and the transferred embryos later lost from the reproductive tract.

Litter birth dates are determined by the gestational stage of the recipient mouse, not by the stage of development of the embryos returned to the uterus. In the manipulation of one-cell embryos, some prefer to return them to recipients plugged on the same day, based on the fact that in vitro culture conditions are never as good as in vivo conditions. Some prefer to culture such embryos to the two-cell stage and then implant into day 0.5 p.c. recipients. This can be useful where you may wish to know that the procedure applied to the embryos has not proved developmentally harmful. Another topic for discussion and dissent would be whether to implant embryos into both horns, or just one. When implanting into oviducts, our preference would be to implant embryos of any stage unilaterally, although bilateral implants work just as well. Unilateral implants are less stressful for the animal, since the extent of the surgery is reduced. It is also less work for the operator if 20 early stage embryos are to be implanted in one oviduct, rather than divided between two. In addition, litter sizes do not seem to be reduced by the use of unilateral implants. When implanting one- or two-cell embryos bilaterally, we recommend implanting 8–15 embryos per oviduct, depending upon the strain of implanted embryos – more for inbred strains and fewer for outbred or hybrid strains. If implanting one- or two-cell embryos unilaterally, we recommend implanting between 20 and 30 embryos per oviduct.

E3.5-stage embryos may be implanted bilaterally or unilaterally. At this stage, embryos will only implant in the manipulated uterine horn, because the movement of embryos to the unmanipulated horn is less likely in the few hours before physiological implantation into the uterine wall begins. Thus, under these conditions, one should reduce the number of embryos transferred to a horn. When implanting blastocysts bilaterally, we recommend seven embryos per horn. If implanting unilaterally, we recommend implanting seven to ten embryos only in one horn.

11.4.2 Oviduct Transfers in Mice: Unilateral, Infundibulum or Ampulla

Embryos at any stage of development, up to blastocyst stage at E3.5, may be successfully implanted into the oviduct of day 0.5 p.c. females. The embryos will remain viable and in a suspended state of development until the uterus is ready for implantation. However, whether you implant blastocyst stage embryos or one-cell stage embryos, if you are using day 0.5 p.c. recipients, embryos must be deposited into the oviduct of recipient females. If implanted directly into the uterus of day 0.5 p.c. recipient mice, the embryos will be lost from the reproductive tract before they are able to implant. In selecting suitable female mice for the implantation of embryos at E0.5, it is important to use females identified as having mated the previous night, as indicated by the presence of a copulatory plug on the morning of the day of implantation. Also, once the female is prepared for implantation, you should be able to identify the presence of corpora lutea on the surface of the ovaries (Fig. 11.2), plus the presence of the ampulla, holding ova, in the upper part of the oviduct (Fig. 11.3). If one or both of these features are not present, it is advisable not to use that female as a recipient, but to move on to the next potential recipient female. In our experience, up to 10% of plugged and theoretically pseudopregnant females do not exhibit one or more of the features detailed above. To compensate for this, we suggest (see Sect. 11.2.2.3) that you plan to produce more recipient females than you anticipate needing for use.

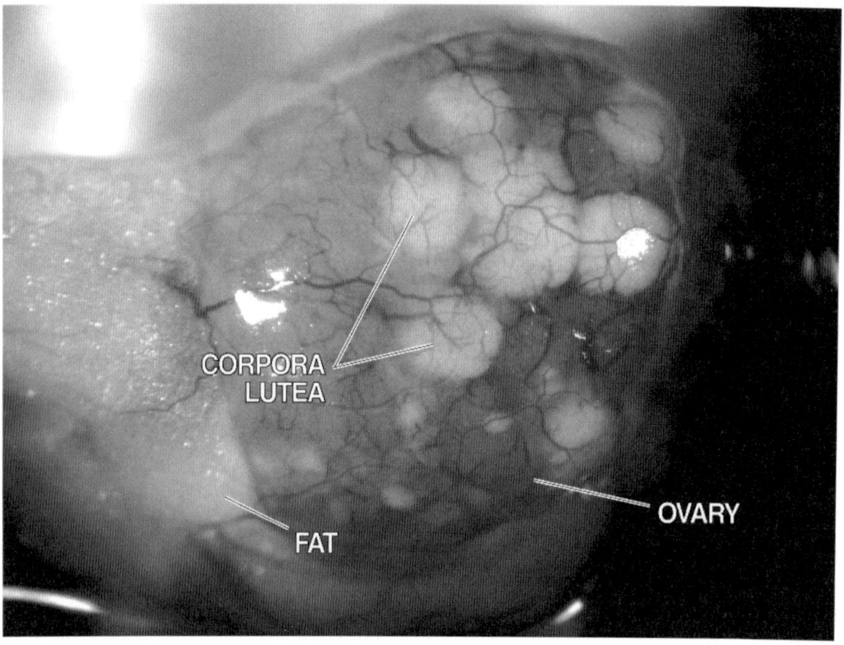

Fig. 11.2. Photograph of mouse ovary, corpora lutea

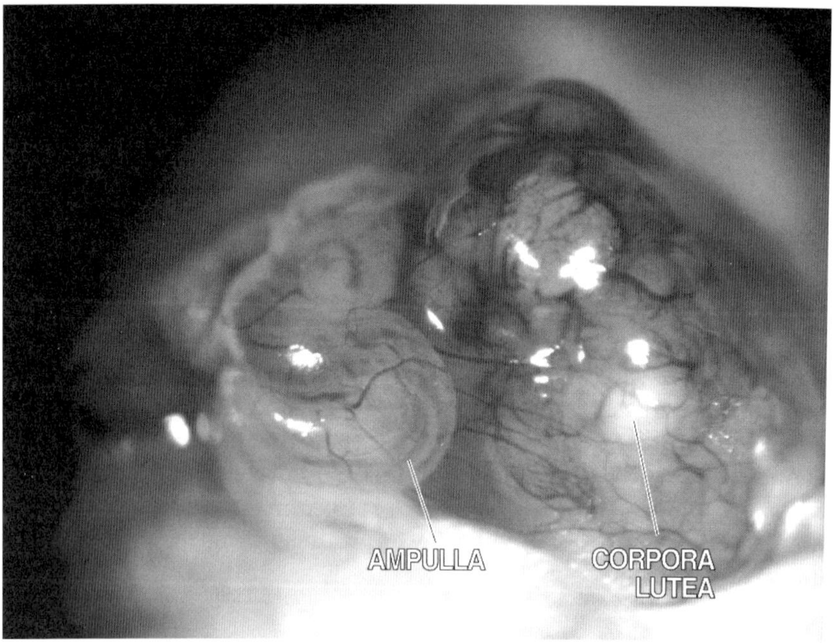

Fig. 11.3. Photograph of mouse ovary and ampulla

11.4.2.1
Equipment Needed

- Two dissecting scopes, one with transmitted light base, one with reflected light source.
- Scissors, 11 cm, straight, fine pointed
- Forceps, 11 cm, serrated, straight
- Forceps, 11 cm, serrated, curved
- Forceps, 11 cm, watchmakers no. 5 forceps, straight, fine pointed
- Serrafine forceps, 3.5 cm
- Acupuncture needles, 2.3 cm, 32 g, or hypodermic needle, 30 g
- Mouth pipette and hand-pulled embryo transfer pipettes
- Drummond Dialamatic, optional (see Sect. 3.4)
- Fur clippers
- Optical petroleum jelly
- Skin disinfectant (iodine and 70% ethanol)
- M2 medium
- Plastic Petri dish, Falcon bacterial dishes, 35 mm
- Suitable anesthetic (e.g., Ketamin:xylazine mix – 90 mg/kg and 10 mg/kg respectively)
- Analgesic, such as Ketaprophen given subcutaneously or Bupivicane applied to the musculature at the incision site

11.4.2.2
Procedure

Prior to preparation of the recipient mouse, set up the embryos to be implanted. Retrieve them from the incubator and place in 2 ml M2 in a 35 mm Petri dish, on the microscope with a transmitted light base. Set the other dissecting scope up for use during surgery somewhere nearby, since you will travel between the two microscopes while implanting embryos.

1. Anaesthetize the recipient mouse.
2. Lubricate cornea on both eyes, to prevent damage caused by drying while anaesthetized.

3. Shave the back of the animal and prepare skin for surgery by swabbing with iodine and 70% alcohol alternately, three times each.

4. Check that the animal is anaesthetized to a deep enough level for abdominal surgery, by applying pressure to one hind foot. If the anesthesia is not deep enough, then she will move her foot away from the pressure. In this case, deepen the anesthesia to an appropriate level, before starting surgery.

5. Using aseptic technique, make a single incision, about 2.5 cm long, centrally, midway down the back of the animal.

6. By blunt dissection, separate the skin from the membrane on the surface of the abdominal musculature, i.e., lift the skin at the incision with forceps and insert closed scissors pointing down toward the bench or work surface. Open the scissors in such away that the blades move between the underside of the skin and the abdominal musculature.

7. The fat pads associated with each ovary can be seen through the musculature of the abdominal wall. They are located midway between the lowest rib and the pelvis, on each of left and right sides. Move the skin incision over the left ovarian fat pad. Make an incision in the musculature over the site of the left ovarian fat pad.

8. With curved and serrated blunt forceps, take hold of the fat pad and lift it out of the abdominal cavity. Clasp the fat pad with a pair of serrafine forceps and lay the forceps down, such that the ovary, oviduct and upper part of the uterine horn are resting on the surface of the back of the animal. Locate the infundibulum and, according to its position, place the animal such that you could comfortably slide an embryo transfer pipette into the infundibular opening. The infundibulum is the open end of the oviduct and can usually be found nestling close to the ovary, covered by the bursa (the membrane surrounding the ovary). The infundibulum appears a little like a doughnut ring – the striations of the inner surface appear to be rolled back in a thickening at the end of the oviduct (see Fig. 11.4).

9. Pick up in the transfer pipette an appropriate number of embryos (as given in Sect. 11.4.1). Place an air bubble in the pipette before and after the group of embryos to be implanted and draw them up into the pipette with as little medium as possible.

10. Hold a pair of fine forceps in your left hand and the transfer pipette in your right, together with a hypodermic needle of 30 g size. Lift the bursa away from the ovary and oviduct, in the region of the infundibulum.

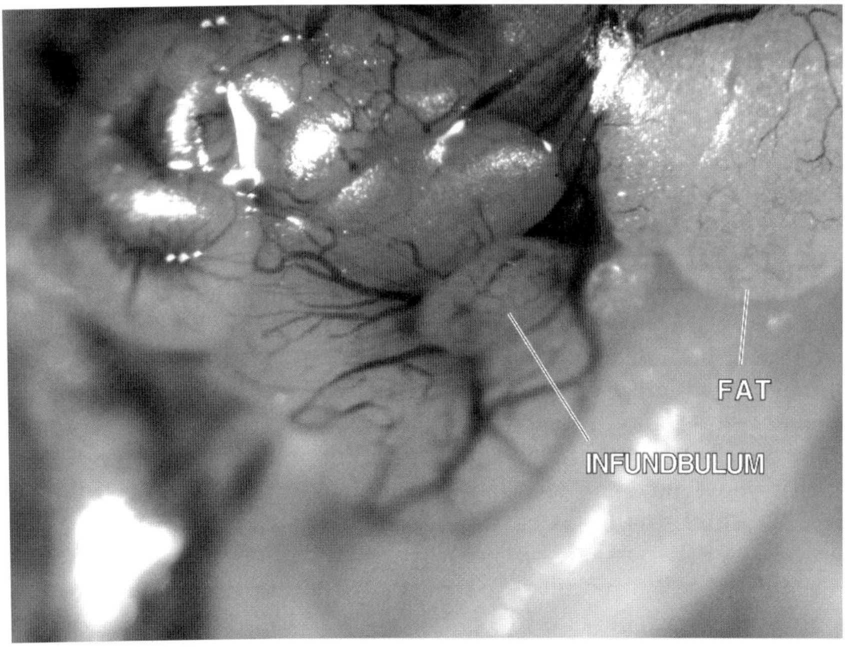

Fig. 11.4. Photograph of mouse oviduct, ampulla and infundibulum

Using the hypodermic needle make a small cut in the bursa, avoiding blood vessels such that you can insert the tip of the transfer pipette into the infundibulum via the cut in the bursa. The embryos can then be deposited in the oviduct. Look for the presence of two air bubbles in the oviduct, an indication that the embryos between the bubbles have been deposited in the oviduct.

11. Withdraw the pipette and release the fat pad from the serrafine forceps. Using a pair of fine forceps to lift the abdominal muscle wall incision and the curved blunt forceps to manipulate the ovarian fat pad, return the uterus, oviduct and ovary to the abdominal cavity. Try to do so by handling the fat pad only. Then close the wound in the skin with skin clips.

12. Keep the animal warm until recovery from the anesthetic is complete. Litter size may be influenced by post surgical body temperature of recipient animals (Bagis et al. 2004).

Note: The incision in the abdominal wall is to one or each side of the incision in the skin. This, and the fact that these incisions are on the dorsal surface of the animal, renders it unnecessary to close the muscle wall with suture. Nevertheless, this is an option and may be required by institutional policy. In this case, use a sterile synthetic absorbable suture, 4/0 or 5/0.

If bleeding is a problem in puncturing the bursa, then a few drops of the vasoconstrictor, epinephrine, can be used in a 1% solution, prior to puncture.

The above procedure describes a unilateral implant. This has been seen to be as efficient as bi-lateral implants and has the advantage of being quicker to complete, thus benefiting both the operator and the recipient animal. One alternative to implanting into the infundibulum would be to implant directly into the ampulla, which is an expansion in the oviduct, just distal to the infundibulum (see Fig. 11.3). In this case, a small hole would be placed in the ampulla with an acupuncture needle or a 30 g hypodermic needle, and the embryo transfer pipette placed into the ampulla directly, for the deposit of embryos into the ampulla. This has the advantage that you avoid the potential for a problem with bleeding from the bursa. Also, the ampulla is easier to locate than the infundibulum. The disadvantage may be that some embryos may be lost from the ampulla as, or after, embryos are deposited, particularly if medium volume is not kept to a minimum, because it is possible for implanted embryos to wash back out through the hole made for the implantation pipette. This should not be confused with the loss of endogenous unfertilized eggs. These can sometimes be seen to be displaced by the manipulated embryos as they are transferred into the ampulla. These unfertilized eggs may be visible because they are still enclosed by cumulus mass cells. Sometimes they can be seen to emerge from the ampulla via the hole made for the implant pipette as manipulated embryos are deposited. They emerge as a soft mass and may cling to the implant pipette. Another alternative route for implantation of embryos at this stage would be to make a small hole in the oviduct, between the infundibulum and the ampulla, outside of the bursa. (See Chap. 4, *Cloning the Laboratory Mouse by Nuclear Transfer*, Sect. 4.3.13.)

11.4.3
Uterine Transfers in Mice

This procedure is for the transfer of blastocysts and late stage morulae only. Recipient females will have been identified as having mated by the presence of a vaginal plug on the morning of day 0.5 p.c. These females will have been separated from the males and group housed until day 2.5 p.c., on which day embryos at E3.5 will be implanted. In this case, the presence of Graafian follicles on the ovaries of recipient females should be identifiable (Fig. 11.5).

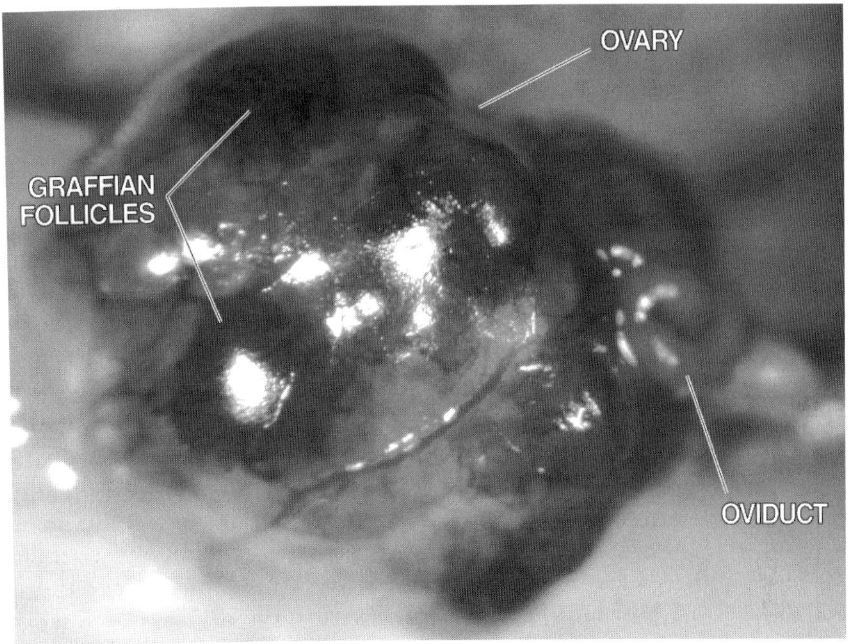

Fig. 11.5. Photograph of mouse ovary and Graafian follicles

11.4.3.1 Procedure

Work through oviduct transfer steps 1–8 (see Sect. 11.4.2.2)

9. Pick up seven blastocysts in an embryo transfer pipette.
10. Hold a pair of fine forceps in your left hand and the implant pipette and a 26. g acupuncture needle, or a 25 g hypodermic needle, in your right. Placing your left hand around the top of the animal and using the fine forceps, pick up the uterine horn at the end, near the oviduct. Put a little tension on the horn by moving it towards your right hand – the small amount of tension put on the tissue of the uterine horn will make it easier to make a hole in the wall of the uterus.
11. With the needle, make a hole in the wall of the uterus, just below the point where the forceps grasp the horn. Be sure that the needle goes into the lumen of the horn and does not remain or become embedded in the endometrial layer of the uterine wall.
12. Withdraw the needle and place the embryo transfer pipette in the hole made by the needle. Deposit the blastocysts in the lumen of the uterus by expelling them from the transfer pipette. If it seems difficult to deposit

the embryos, in that you cannot expel them from the transfer pipette, then adjust the position of the tip of the pipette by moving it deeper into the lumen of the uterus or withdrawing it a little. This may free the open end of the pipette in the event that it is pressed against, or lodged within, the wall of the uterus.

13. Withdraw the transfer pipette and return the uterus to the abdominal cavity, as described above. Repeat the procedure to deposit seven blastocysts in the right uterine horn. You may find the logistics easier to deal with if you turn the recipient mouse around, so that you are able to work comfortably.
14. Return the second uterine horn to the abdominal cavity by manipulating the ovarian fat pad. Close the skin incision with skin clips.
15. Keep the animal warm until recovery from the anesthetic is complete.

11.4.4 Synchronization and Embryo Transfer in Recipient Rats

Even though the embryos may be cultured in vitro to the two-cell stage in KSOM, M16 or equivalent (Charreau et al. 1996), we recommend that you avoid keeping them in culture unless mR1ECM medium is available (Miyoshi et al. 1995, 1997; Zhou et al. 2003). Rather, implant them on the same day as injection or manipulation. In this case, the mating of putative recipient rats should occur on the same day as the mating of embryo donor rats. You may be working with vasectomized males and virgin females in the generation of pseudopregnant recipient rats, in which case, the numbers needed are similar to that of mice. Assume a maximum of 20% of paired females will be in oestrus and will mate during one night. Thus, mate five or six times the number of recipients needed and mate them at a ratio of two females to one male. Alternatively, select a smaller number of females for mating by using those females with external signs of being in oestrus or which demonstrate the presence of epithelial cells in vaginal smears (Rough 1993). Mate a similar smaller number of females that have been synchronized by an IP injection of 20 IU hCG, on day –3.5 p.c. (Mullins et al. 2003). If possible, house these animals at mating on grid bottom cages, with paper on the waste collection tray. This is because the post-coital plug more readily shrinks and falls out in rats than in mice. If you house these rats in solid bottomed cages on typical rodent bedding, you will quite likely lose and/or miss the plug(s). Post-coital plugs in rats are more difficult to see than those in mice, because they are generally situated more deeply in the vagina than in mice.

A third approach to making pseudopregnant recipient female rats available is to purchase timed mated females, such that females are due to litter 1–2 days before the injection day. Post-parturition females can then be paired with vasectomized males on the night prior to injection. We have found that Sprague Dawley rats have good mothering ability and make suitable recipient animals. So, for example, if we are injecting on a Friday, we purchase animals that were plugged, and therefore at day 0.5 p.c., on Wednesday and on Thursday, 22 and 23 days, respectively, before the injection day. The goal is to obtain females that will litter down on the day or the night before the planned injection day. In this way, you can take advantage of post-partum oestrus in these females and thus have a small number of animals that will be ready for mating and ready to receive transferred embryos on the following injection day. We purchase females that were each mated on one of two nights because the period of gestation may vary a little between individuals of any particular strain. However, with enough experience with your strain of choice, it should be possible to narrow this down to one optimal plug day. In the rat, post-partum oestrus starts 10–20 h after parturition and lasts for 12 h. Thus it is likely that some of these females will be in postpartum oestrus on the night of Thursday, the day before injection. This being the case, normally six postpartum females mated with six vasectomized males are enough to generate three or four recipient females on 1 day. It is possible to return the implanted recipient female to her pups and allow her to rear them over the next 3 weeks.

As with mice, it is possible also to transfer embryos into recipient female rats that have mated with non-vasectomized males. In this case, expect to obtain a litter of pups containing injected and non-injected animals. This approach is useful only when working with a high efficiency technique, such as lentiviral injections or simple embryo transfer for re-derivation purposes. In the generation of genetically altered rats by pronuclear injection, for example, a great deal of additional microinjection work would be required to compensate for the reduced number of putative founder animals likely to be present in each litter born.

11.4.5 Oviduct Implants in Rats, by Infundibulum or Ampulla

11.4.5.1 Equipment Needed

- As for mice
- Add scissors, 12 cm, straight, pointed
- Use larger serrafine forceps, 6.0 cm in length

- Also larger acupuncture needles, 5 cm in length and 26 g, or hypodermic needle, 25 g

11.4.5.2
Procedure

Follow the procedure as for mice. There are some differences in the anatomy of rats as compared with mice. The bursa is more highly vascularized in rats than in mice. This may vary from strain to strain, but has been our experience in working with Sprague Dawley and Fischer rats. In addition, the abdominal muscle wall of recipient rats is disproportionately thicker in rats than in mice and may be considered a more compelling case for suturing after surgical transfer of embryos, before closing the skin.

You may be working with recipient females in which a plug was not found on the morning after mating. The plug may have been lost or the female may not have mated at all. In this case, the only approach would be to evaluate the suitability of each female as a recipient based upon the appearance of the reproductive tract. In suitable females, the corpora lutea should be clearly visible upon the surface of the ovaries of the recipient female. The corpora lutea on the surface of rat ovaries are more exaggerated than in mice, the ovary taking on an appearance similar to a small bunch of grapes (Fig. 11.6). Also look for a swollen ampulla. The presence of both these conditions is the best indicator that this female is ready to receive one-cell embryos.

11.5
Murine ES Cells

11.5.1
Commonly used ES Cell Lines

There are few commercial sources of ES cells. Those that exist at the time of publication are available through Lexicon (The Woodlands, TX; www.lexicon-genetics.com), Primogenix (St. Louis, MO), Specialty Media (Phillipsburg, NJ; www.specialtymedia.com) and ATCC (Manassas, VA; www.atcc.org). Of stem cell lines available, each may have its own characteristics (see Table 11.4). ES cells are best used at early passage because the cells may acquire spontaneously arising genetic mutations when kept in culture. As a guide, try to obtain cells for experimental use at passage ten or below. If you wish to establish a bank of cells for use in multiple experiments, then try to obtain a vial at about passage five, so that you can expand the cells and freeze them at passage seven to ten for use in

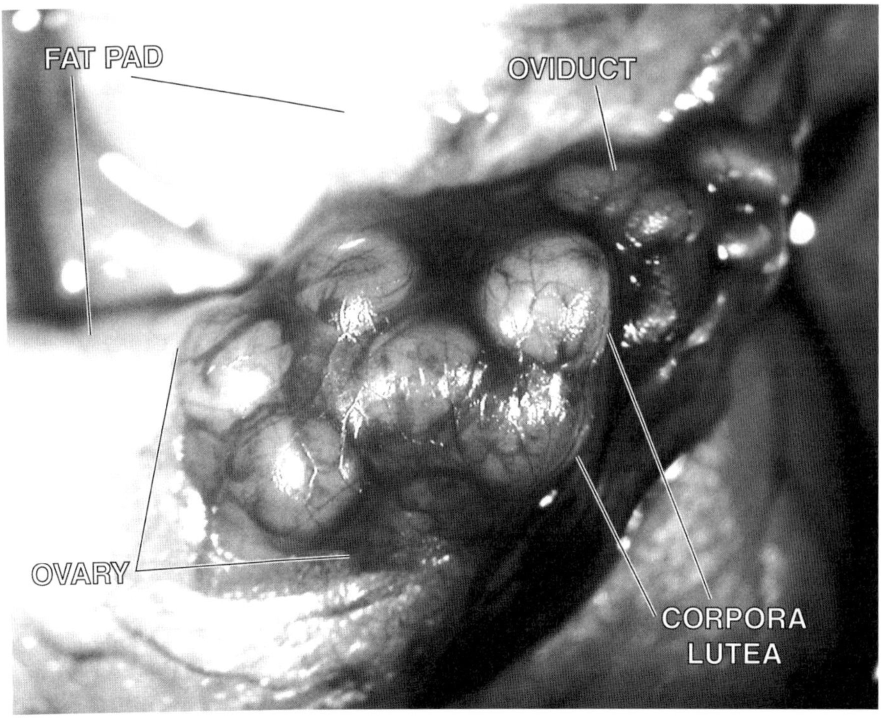

Fig. 11.6. Photograph of rat ovary and corpora lutea. Note approximately 50% of ovary obscured by fat pad

individual experiments. The ultimate solution would be to develop one's own ES cell line, by culturing from blastocysts. This is ideal because you will have first-hand access to the lowest passage cells possible. Also, it has been reported that ES cells perform best if kept in consistent culture conditions. Any change in culture conditions, such as might occur by transfer to another laboratory, has the potential to reduce the pleuripotency of ES cells. For example, differences in water quality used to make media, as well as differences in the source and condition of supporting fibroblast cell lines, may reduce the capacity of the cells to transmit through the germline. Even a switch between two otherwise superb feeder cell lines has been purported to reduce ES cell potential.

The culture and use of ES cells is a specialized field and it is not our goal to include that information here. Further detailed reading can be found in several recommended texts (Wurst and Joyner 1995; Torres and Kuhn 1997; Abbondanzo et al. 1993; Doetschman 1994; Robertson 1987; Nagy et al. 2003). Most ES cells available today originate from one of several substrains of the 129 inbred mouse. Because of a pre-disposition towards

Table 11.4. Commonly used embryonic stem cell lines

Embryonic stem cell line	Originator	Parental mouse strain	Genes for coat color at A, B, C and P loci	Characteristics
EKCCE	M. Evans	129SvEv	AABBCC	
E14	E. Robertson	129/Ola	$AABBc^{ch}cch_{pp}$	
HM1	D. Melton	129/Ola	$AABBc^{ch}cch_{pp}$	Feeder free
AB-1 and AB-2	A. Bradley	129SvEvBrd	AABBCC	Hprt deficient
J1	R. Jaenisch	129SvJae	AABBCC	
D3	T. Doetchman	129SvPas	AABBCC	
R1	R. Jaenisch	129SvJ x 129Sv-+p	$AABBCc^{ch}P_{p}$	Aggregation chimera competent
See Table 3.1B, Chap. 3, this manual	R. Jaenisch	B6129	AaBBCC	Tetraploid embryo complementation competent
Bruce4	C. Stewart	C57BL/6	aaBBCC	

naturally occurring teratocarcinomas, the 129 mouse strain facilitated the establishment of embryonal carcinoma (EC) cell lines – a cell type morphologically similar to ES cells and capable of differentiating into various cell types under specific culture conditions; albeit in a more limited capacity than ES cells. Despite their more restricted use, research on 129-derived EC cell lines paved the way for the development of ES cell technology, now commonly used in gene targeting experiments in mice.

Currently, many of the most widely used pleuripotent ES cell lines available are derived from strains of the mouse line 129. However, the C57BL/6 mouse strain is the best characterized scientifically, and therefore many researchers consider it necessary to backcross their mouse model onto a C57BL/6 background. This can take up to 2 years, since it is necessary to go though ten generations of backcrossing before reaching a genetic background closely approximating to C57BL/6. More recently, C57BL/6 ES cells have become available and although they do not perform with the same efficiency as 129 derived lines (Seong et al. 2004), there is an obvious advantage in putting a mutation directly into this genetic background. In addition, various hybrid ES cells derived from matings between either two different 129 substrains, or from two completely different inbred strains now exist. These hybrid cell lines are proving to be capable of giving rise to animals that are wholly of ES cell origin, in the founder generation (see Chap. 3, *Generation of ES Cell-Derived Embryos and Mice* by Tetraploid-Embryo Complementation by K. Eggan and R. Jaenisch).

11.5.1.1
Coat Color Marker Genes for Chimera Production

When selecting a strain of mouse to use as an embryo donor, it is necessary to consider (1) whether the ES cells will be able to colonize well in that strain, and thus contribute towards the production of germline transmitting chimeras; (2) whether the coat color genes are such that gametes arising from both the ES cell component and the host blastocyst will be distinguishable in the F1 generation, and (3) which of these candidate strains will respond well to hormone priming to produce the maximum number of embryos for use. The genome of the host blastocyst will not contribute to the background of the mutation, since gametes arising only from the portion of the gonads that have been colonized by ES cells will be selected for in the F1 generation.

Typically, 129 ES cells (AABBCC at these three loci for coat color) are injected into C57BL/6 (aaBBCC) host blastocysts. The agouti coat color of the 129 mouse present in ES cells can be used as a marker to determine the extent of chimerism in resulting animals, when compared against the black coat color of the host blastocyst. In addition to C57BL/6, there are other strains that might be used as hosts for 129 ES cells with varying degrees of success. For example, the outbred CD-1 strain can be used as a host blastocyst with 129 ES cells. However, usually only low-percentage chimeras are obtained as 129 ES cells appear to colonize CD-1 blastocysts poorly, perhaps due to growth competition with host cells in the developing embryos. On the other hand, some hybrid 129 substrain ES cell lines, such as R1, appear to be more vigorous, and are able to adequately colonize in a CD-1 host blastocyst environment. Chapter 3 in this volume addresses these principles. The ability of ES cells to colonize in embryos from an outbred source may alternatively be a property of low passage ES cells, rather than a question of "hybrid vigor" in ES cells (Auerbach et al. 2000).

When C57BL/6 ES cells (aaBBCC) are used, several albino mouse strains are good candidates for the supply of host blastocysts. These include BALB/c (AAbbcc), FVB/N (AABBcc), and the more recently developed strain C57BL/6Tyr(c-2J) (aaBBcc), now available from JAX (www.jax.org). C57BL/6Tyr(c-2J) arose from a spontaneous mutation that occurred in the tyrosinase gene in an animal within the C57BL/6 colony at the Jackson Laboratory (Bar Harbor, ME; Jax Mice Notes 2003). C57BL/6 ES cells have been reported to produce germline transmitting chimeras at higher frequency when injected into co-isogenic blastocysts of the C57BL/6Tyr(c-2J) strain than into other albino strains (Schuster-Gossler et al. 2001; Seong et al. 2004). Although offering great hope as a source of host blastocyts for experiments involving C57BL/6 ES cells, it is not yet known how well this strain responds to hormone priming relative to other strains.

11.5.1.2 Coat Color Indicators for Germline Transmission

Somatic cells within the mouse are diploid and therefore carry two identical copies of each gene, including those controlling coat color. Gametes, on the other hand, having undergone meiosis, are haploid and therefore carry only a single allele for each coat color gene, which they contribute to the zygote at fertilization. The specific coat color allele carried by a gamete produced by a chimeric animal, whose tissues are composed of two populations of cells harboring separate and distinguishable genes for coat color at some loci, directly reflects the portion of the gonad from which the gamete arose in the chimera. If derived from the portion colonized by the injected ES cells, then the individual coat color alleles will be of ES cell origin. If derived from the portion colonized by the host blastocyst, then the individual coat color alleles will be of host embryo origin. Thus, with the appropriate selection of mating partners for the chimeras produced, dominant coat color markers can be used to select F1 offspring (F0 being the founder chimeric animals) in order to determine if the genetic modification is heritable and to identify offspring as candidate carriers of such modifications.

When chimeric mice produced from 129 ES cells (AABBCC; carrying the dominant agouti allele) and C57BL/6 host embryos (aaBBCC; carrying the recessive agouti allele), are mated to C57BL/6 mice, agouti offspring (AaBBCC) are produced only by the contribution of the dominant agouti allele from the 129 ES cells. When C57BL/6 ES cells (aaBBCC) are used, the dominant forms of the B and C genes for black coat color may be used as a marker for germline transmission, in combination with blastocysts from albino strains (Lemckert et al. 1995; Auerbach et al. 2000). Chimeric mice produced from this combination will produce gametes containing either the dominant "C" form at the C locus (derived from the ES cell) or the recessive form of the gene for color, "c" (derived from the host embryo). Therefore when these chimeras are mated with an albino mouse (homozygous for the recessive form of the C gene), F1 offspring will be either albino or colored (color depends upon the particular albino strain used). Non-albino mice will contain genetic material from the C57BL/6 ES cells. The best choice for mating C57BL/6 ES cell chimeras is the albino C57 mouse – C57BL/6Tyr(c-2J; see below). Here, coat color genes for ES cells are aBC, and gametes from the host blastocyst, whether BALB/c, FVB/N or C57BL/6Tyr(c-2J), will be homozygous at the C locus for the recessive form of the gene. Gametes from mating females, C57BL/6Tyr(c-2J), are all aBc. So, again, any animal expressing color, i.e., that is not albino, will be carrying a haploid set of ES cell chromosomes. In this case, such animals will be black. In addition, a great advantage in using this strain is that F1 offspring may be considered "inbred", due to the co-isogenicity of C57BL/6 and C57BL/6Tyr(c-2J) strains. (Note: some albino strains, such as BALB/c and FVB/N, carry the dominant agouti allele. Where host blastocysts were derived from one of these strains, the agouti gene can be used to select offspring that have *not* arisen from the ES cell component of the testes.)

Therefore, just as it is important to select an appropriate strain for use as host embryo in order to determine the degree of chimerism, it is also important to select the appropriate strain in which to mate chimeric animals in order to determine if the ES cells used can colonize the germline. Each haploid set of chromosomes must be distinguishable from the other in combination with a haploid set of chromosomes from the mating partner. In most settings it is standard to mate chimeras to the strain used for blastocyst supply. One should also consider onto which genetic background this mouse model should be established for the generation of experimental animals. It should be noted that only 50% of F1 mice determined to be of ES cell descent will contain the gene modification (assuming only one allele was modified during the in vitro stages of the experiment) due to segregation during meiosis. Therefore, F1 offspring of the desired coat color must be further screened for the presence of the gene modification. In subsequent generations, independent assortment and random segregation of chromosomes at gametogenesis renders the coat color indicator irrelevant.

Where F1 ES cells have been used, other coat color genes in a recessive form may become evident in F2 generations and onwards. The R1 ES cell line, for example, is homozygous at the agouti and black loci, for dominant forms of each gene. But it also carries one copy of the c^{ch} (chinchilla) gene, at the C locus and one copy of the recessive gene p (pink eye) at the P locus. Chimeric animals appear black agouti in combination with the host blastocyst coat color, and the dominant agouti gene may be used to identify ES cell germline F1 animals. However, at an intercross between two F1 animals, recessive copies of the genes for chinchilla and/or pink eye may come together. Thus, these phenotypes may be identifiable in F2 offspring and subsequent generations.

11.5.2 Elementary Karyotyping

When carefully cultured under optimal conditions, ES cells are pleuripotent cells that have the ability to contribute to all cell lineages of the developing embryo. Genetically altered mice (knock-out, knock-in, point mutations) are derived by the injection of recombinant ES cell clones into blastocyst stage embryos. However, we know that these cells are not wholly chromosomally stable in culture, though some ES cell lines are more stable than others. ES cells used to generate such mutations have undergone electroporation and selection for homologous recombination at a specific site for the gene mutation. In order to make this alteration, ES cells must be grown in culture for five to seven passages, in addition to routine passages accumulated during the time the ES cell line was initially expanded. Typically, manipulated ES cells are at passage 16 or higher before being returned to the embryonic environment.

Recent findings show that chromosome abnormalities in ES cell clones are a rather frequent occurrence (Liu et al. 1997; Longo et al. 1997). Abnormal chromosome counts of between 38 and 42 chromosomes are most usual, the normal number for mice being 40. One of these abnormalities, trisomy of chromosome 8, has been seen to confer a growth advantage to cells in culture (Liu et al. 1997) When these particular cells are returned to the embryonic environment, they appear unable to contribute to the production of gametes, although they are able to contribute to somatic cell lineages, thus good chimeric animals can be produced with such ES cells. Conversely, the slower growing ES cell clones are, for the most part, euploid and able to contribute to the germline. Clones with a predominantly normal chromosomal number must be used if germline transmission is to be expected from animals generated by these means. Many established ES cells lines are male, i.e., XY, because of the advantage conferred by the use of males in test mating putative germline transmitting chimeras. Another quite common abnormality would be the loss of the Y chromosome, giving rise to a count of 39 chromosomes. In this case, when injected into female blastocysts, female chimeras capable of germline transmission may be generated. XY cells present in the gonads of female animals are incapable of giving rise to viable ova. However, XO cells are capable of generating viable ova.

We have found it beneficial, when working with an ES cell line from an inbred strain of mouse, to screen recombinant clones by gross karyotyping prior to injection. Only clones that are 55% or more of normal karyotype will be injected. Prior to pre-screening of clones for injection, we found that only 50% of injected clones would result in the production of germline transmitting chimeras. In our hands, 50% of the clones derived from CJ7 ES cells contain more than 50% of aneuploid cells. Thus, by pre-screening clones by chromosome count, we have been able to raise to 100% the frequency of germline transmission from chimeras derived from ES cell clones (with the infrequent exception of targeted mutations that we think have reduced the pleuripotency of the cells).

11.5.2.1
Preparation of Chromosome Spreads: Tools and Equipment

- ES cells – 35 mm plate at 80% confluence works well, thus having been 1 or 2 days in culture only. (Only dividing cells will provide spreads for counting. Therefore inactivated feeder cells will not skew results. However, if this is a concern, plate ES cells out on gelatinized plastic alone.)
- PBS
- Trypsin EDTA (Gibco, Grand Island, NY; cat. no. 25200-072, 0.25%, 1 mM EDTA-4)
- Colcemid (Gibco 15210–040, stock solution at 10 μg/ml)
- Potassium chloride (KCl), 0.56% (0.075 M)
- Methanol:glacial acetic acid (3:1) – freshly prepared each time. Keep on ice, use from ice.
- Microscope slides, cleaned with 70% ethanol
- Coverslips
- Vectashield mounting medium with DAPI, (H-1200, Vector Laboratories, Burlingame, CA)

11.5.2.2
Protocol

1. Passage a 35 mm plate containing ES cells to be karyotyped (onto a gelatinized, feeder-free dish), 1 day prior to preparing spreads, to ensure actively dividing cells are present. In this case, passage 50% of an 80% confluent 35 mm dish onto a gelatinized 35 mm dish. (We recommend using a feeder-free dish because the presence of feeder cells reduces the percentage of cells that will be seen to be in metaphase in any one prep. But using a feeder-free dish is not essential.)
2. Feed cells 2–3 h prior to adding Colcemid. Add Colcemid to a concentration of 0.1 μg/ml on the ES cell dish, for 1 h at 37 °C. (Add 20 μl stock solution to 2 ml medium.) Colcemid will arrest dividing cells in metaphase.
3. Aspirate medium and rinse with PBS. Trypsinize the cells. (Important to trypsinize well, for 10–15 min, to ensure a single cell suspension.) Stop the action of trypsin by adding an equal volume of ES cell medium.
4. Spin down for 10 min at 900 rpm.

5. Pour off supernatant and re-suspend pellet in remaining liquid. Add 5 ml 0.56% KCl. Let stand at room temperature for 10 min. The addition of this hypotonic solution causes the cells to swell. At a mid-point during this 10 min incubation period, invert the tube gently, once only, to re-suspend the settled cells in the KCl solution. Optimal incubation time in this solution may vary. Too long an incubation time will cause chromosomes to spread everywhere when the spreads are prepared, and it becomes difficult to determine the origin of each chromosome; too short, and chromosomes will not separate well enough for clear counting.
6. Spin down for 5 min at 900 rpm.
7. Pour off supernatant and resuspend pellet in remaining liquid. It is extremely important that the pellet be resuspended completely before adding the fixative. Using freshly prepared fixative, add 5 ml slowly to the cell suspension. Add the first 2 ml dropwise, tapping the tube on the bottom after each addition to mix. Add the remaining 3 ml at a faster rate, still making sure to mix well. Let stand for 10 min at room temperature.
8. Spin down for 5 min at 900 rpm.
9. Pour off supernatant and resuspend pellet. Add fixative as in step 7.
10. Spin down for 5 min at 900 rpm.
11. Pour off supernatant and resuspend pellet in a small amount of fixative, usually less than 0.5 ml for cells harvested from a 35 mm dish. The suspension should be slightly cloudy. Continue to keep the suspension on ice until used for step 12.
12. Let one drop of the cell suspension in fixative fall onto a clean slide, from a height of about 10 cm above the slide, using a P-200 Pipetman and yellow tips. Allow to air dry. View under a microscope to determine if there are too few or too many cells (the latter being the more likely problem). If too few, spin down again and resuspend in a smaller volume. If too many, add a small amount of fixative to the suspension and drop again onto a clean slide. The cells should appear fairly well separated, with clumps the size of only two or three cells.
13. Apply one drop of mounting medium with DAPI to the spread and overlay with a coverslip. You can keep the slides for up to 14 days at 4 °C
14. Photograph the chromosome spreads using a fluorescent microscope with a 100 × objective. A regular 35 mm camera works well, but we pre-

fer to use a digital camera. This is because when the images are downloaded onto a disc and then reproduced on a computer monitor, the image is slightly enlarged. This simplifies counting of spreads a little. We recommend counting 24–36 spreads per clone. *Note:*

- The use of a more concentrated cell suspension does not significantly increase the number of spreads per slide.
- Warming of slides to 40 °C in a water bath prior to dropping cell suspension does help.
- In any prep, it is likely only about 1 in 50 cells will be in metaphase and nicely spread on the slide.
- In step 5, inverting the tube once during KCl incubation appears to significantly increase the number of good spreads finally produced.

11.6 The Establishment of Stable Strains from Genetically Altered Animals

11.6.1 Breeding from Chimeric Mice

Chimeric animals arising from the injection of manipulated ES cells are recognized as founders in the establishment of genetically altered strains by this method. Their offspring, which are the first truly heterozygote animals to carry the engineered mutation, are normally referred to as F1 animals. It is likely that the injection of karyotypically normal ES cell clones from a good ES cell line will give rise to pups, 50% or more of which will be chimeric. These chimeras may range from 5% to 95%, this being an estimate of contribution from ES cells as visible by coat color. In the generation of animals carrying any specific targeted mutation, several clonal ES cell lines will have been injected. Although each clone carries the same targeted mutation, animals arising from the injection of each clone should not be mixed or intercrossed with each other. However, individual animals arising from the injection of one clone may be intercrossed or housed together, regardless of specific parentage, as part of one colony. Fundamentally, offspring should remain clonal in origin.

The genetic background of a strain may influence the degree or manner in which the mutation is expressed. Thus it is important to keep detailed breeding records and quite common to establish each mutation on three different genetic backgrounds – one mixed and one each of two pure inbred backgrounds. Scientifically, much is known about the C57BL/6 strain of mouse. Thus it is usual to backcross the mutation onto this background and also onto the ES cell strain of origin. In the latter case, once proven a germline transmitting animal, a chimera may then be mated to an animal from the ES cell strain of origin. F1 animals from this cross are co-isogenic but not inbred, because they are not the result of 20 generations of continuous brother x sister matings. Thus, in order to satisfy the textbook definition of "inbred", in theory these animals should be backcrossed to the ES cell strain of origin from a properly maintained inbred colony.

Most of the initial work in defining a phenotype for a new mutation will be carried out with animals that are most readily available. These will be embryos or animals null for the mutation and arising from a cross between two F1 animals, which in many cases will be on a mixed genetic background, having arisen from a cross between the ES cell strain of origin and the strain that was used to mate with chimeric animals.

11.6.1.1
Mixed Genetic Background Colony, F0 Matings

At the outset, chimeras must be mated to females from a strain that carries suitable coat color genes and which is on an appropriate genetic background. The reasons for this are detailed in Sect. 11.5.1.1 and 11.5.1.2. In making use of the dominant coat color marker associated with your ES cell line of choice, this step will identify chimeras that will transmit the mutation in their gametes, and establish a population of animals carrying one copy of the mutation. The rate at which germline transmission from each chimera becomes evident may vary considerably. Some chimeras will produce offspring carrying one set of chromosomes from the ES cell component immediately. In this case, 10% or 100% of offspring may be identified as carrying the coat color marker from the ES cell component. Some chimeras may appear not to produce gametes from their ES cell component until the third or fourth litter. Poor chimeras (35%) may exhibit germline transmission as low as 1 in 300 pups. Some high level chimeras at around 95% may not transmit at all. These animals may mate, but not produce offspring, their sterility due to some level of hermaphroditism.

Choose the best chimeras for test mating. If two or three high level chimeras from the injection of each clone are available, then mate only these animals. High level chimeras are those estimated to be 75% to 95% of ES cell origin. If such good chimeras are not available, then mate those that go as low as 35%. In chimeric animals generated from the injection of 129 ES cells, chimeras that are less than 30% of ES cell origin are unlikely to transmit the mutation through the germline, because in these animals it is likely that ES cells have colonized the gonads to a small extent only. C57BL/6 ES cells have been seen to perform differently to 129 ES cells, in both the generation of chimeras and subsequent performance of those chimeras (Seong et al. 2004). In these cases, some low level chimeras do transmit though the germline.

Some ES cell chimeras may prove to be sterile. The most common cause for this, where ES cells that are predominantly karyotypically normal have been injected, is hermaphroditism. Most ES cell lines are male, thus all cells carry the Y chromosome. These cells have been injected into host embryos, half of which are female, or XX, at the blastocyst stage. The population of XX embryos with XY ES cells usually results in conversion of the sex of the embryo. This is why, with a good ES cell line, we find that more male chimeras are born than females. In some cases, the conversion of an XX embryo to male is incomplete. These animals prove to be sterile. Sometimes, upon closer examination, both male and female reproductive organs can be identified within the abdominal cavity. Also, the appearance of the genitals may be ambiguous.

As to female chimeras, the few that may be born from the injection of an XY ES cell line will probably be fertile, but unable to transmit the ES cell component through the germline. This is because any XY cells that may form part of the ovaries are unable to give rise to viable oocytes. Thus all the oocytes produced by these chimeras will be of host blastocyst strain. An exception may be where the manipulated ES cells have lost a Y chromosome and as a result, the cells injected were XO. A female chimera with such cells having colonized the gonads *will* be capable of generating viable gametes. However, this is a relatively uncommon occurrence. Therefore, the mating of female chimeras arising from the use of XY ES cells should be considered only as a last resort.

Amongst offspring carrying one set of ES cell chromosomes in the F1 generation, only 50% will carry the mutation, because the mutation was made on only one chromosome of diploid ES cells. These animals will subsequently be intercrossed for the production of null animals, which will be used for experimentation. As generations of intercrossing go by within this closed colony, the genetic background of each animal will become less and less clearly defined, as segregation and independent assortment of genes occurs at each generation. In addition, there is the possibility that as this progresses, deleterious genes may accumulate within the genome and result in the manifestation of inbreeding depression. For two reasons (reliable, reproducible results and strain viability), the mutation should be backcrossed onto one or more inbred strains, for the benefit of genetic stability, and the clear determination that phenotype is the result of the engineered mutation and not a random event in the background genetics of the line. The genetic background of the ES cell strain of origin (probably

129) would provide a completely inbred background and reliable consistent expression of the mutation. It is also quickly accomplished by mating chimeric animals with the strain of ES cell origin, but in the event that this is a 129 strain of mouse, expect that reproductive performance will not be as good as that of C57BL/6. There are also some physiological and behavioral characteristics that differ between inbred strains. Thus it may be important to verify a phenotype on two different inbred backgrounds, in order to be sure experimental results are not a manifestation of inbred strain characteristics. In addition, the C57BL/6 strain, as mentioned before, is very widely used in many fields of research.

Each generation of animals arising from an intercross of two genetically like animals is given a filial or "F" number. Generations arising from backcrosses to an inbred strain are denoted by an "N" number.

11.6.1.2
Mixed Genetic Background, F1 Matings

Identify animals carrying the mutation by PCR and confirm by Southern hybridization. In subsequent generations, it will not be necessary to confirm by Southern analysis, but it is a sensible precautionary measure to take now, prior to the generation of a large amount of stock based on this first result. The most common goal is to generate animals that are homozygous for the mutation (otherwise described as being null for the wild type gene). Therefore, after identification of animals carrying one copy of the mutation, you may need to expand on the numbers of animals that are heterozygous for the mutation. In this case, mate heterozygotes to non-carrying littermates, or to a genetic background strain of your choice. Alternatively, if enough heterozygotes are produced in the F1 generation, intercross these to produce homozygous null animals. Experimental data gathered from the use of these animals is publishable, but is referred to as an initial characterization only. Phenotype should be confirmed on a genetically well defined inbred background.

11.6.1.3
Mixed Genetic Background, F2 Matings

Continue to intercross heterozygous or homozygous null animals in order to generate animals for experiments. Animals may be mated to any other arising from the same clone of ES cell, regardless of specific parentage. But do not mate together animals that originate from different clones. In theory, each clone is identical to the next, within one targeted mutation experiment. However, this may not be true due to additional spontaneous mutations that may have occurred during ES cell culture.

11.6.1.4
Backcross onto ES Cell Line Strain of Origin, N0 Matings

As soon as the test mating of chimeras is complete, and chimeras that produce gametes carrying the mutation are identified, mate these animals to the strain from which the ES cells were derived. Most commonly, this is one of the 129 substrains. All offspring from this mating will need to be genotyped by PCR or Southern blot, since there will be no coat color markers to use as an indicator in the identification of animals of ES cell origin. Genotype every animal weaned, in order to find animals carrying the mutation. In this case, only one or two heterozygous animals will be needed to mate with 129 mice, for the production of N2 animals. In this colony, animals at N1 may be considered to be wholly of ES cell origin, e.g., wholly on a 129 background, although not necessarily genetically identical to each other for background. The reason for this is that during ES cell culture, it is possible that small genetic changes have occurred spontaneously. This could be considered similar to the kind of genetic drift that may occur in an inbred colony that remains closed for 20 generations or more (see p 58 in Festing 1979). Therefore, if you need to reach the text book definition of "inbred" with this colony, continue to backcross to 129 animals by genotyping offspring and selecting one or two animals for mating with pure 129 animals, at each generation. Continue to N10, at which point, all animals in the colony may be considered truly on an inbred background. Throughout the process of backcrossing, you need only keep one or two breeding pairs at each generation. The goal is to get through the generations to a point where the mutation is on an inbred background, and thus all animals within the colony may be considered genetically identical. At this stage, there will probably be no need to generate large numbers of animals for experimentation. (*Note:* In practice, many investigators disregard the possibility of there being minor genetic differences/changes within this N1 generation resulting from the backcross of a chimera to an animal from the ES cell strain of origin. In principal, these animals do not meet the definition of "inbred" but, in practice, they seem to be regarded as close enough for experimental purposes.) The level of homozygosity can be calculated, since at each backcross the level of heterozygosity within the population decreases by 50%. At mating of F0 animals, assuming these are genetically disparate, N1 progeny are 50% homozygous. N2 offspring are 75% homozygous. N3 are 87.5% homozygous, N4–97.3%, N5 – 98.65%, N6 – 99.33%, N7 – 99.66%, until N10 – 99.95% homozygous. N5 animals are widely accepted as being the equivalent of inbred, in the context of a confirmation of phenotype.

11.6.1.5
Backcrosses Onto C57BL/6, N0 Matings

Again, founders for this colony are the chimeras. The goal is to backcross the mutation onto the C57BL/6 inbred (wild type) background. In establishing this colony, N1 animals can be taken from the F1 progeny of the mixed genetic background colony. These animals are from a cross between a chimera, (specifically a 129 gamete) and a C57BL/6 animal. The animals are essentially 50% of 129 origin and 50% C57BL/6, and therefore a first backcross onto C57BL/6. Take a heterozygous F1 animal and further backcross to a C57 mouse, for the production of N2 animals in the C57BL/6 backcross colony. Continue through to N5 or N10 by backcrossing a heterozygous animal to a C57BL/6 at each generation. It may be convenient to take a male carrying the mutation and mate him to two wild type females at each generation. In this case, be sure to set up a reciprocal cross at least once, (using a wild type male and transgenic females) or the Y chromosome from the mixed background genome will be selected for, through to N10.

11.6.2
Breeding from Lentiviral Founder Animals

In the generation of mouse models by pronuclear injection, each animal produced that is carrying the transgene is different to the next, for several reasons. Firstly, the transgene integrates at random, since this is not a targeted event. Thus integration sites vary from founder to founder. Secondly, copy number may be different, since the transgene may integrate either as a single copy or in tandem arrays at a single location, on one chromosome. Thirdly, in a small percentage of cases, integration may not take place until after the zygote has gone to the two- or four-cell stage. This will result in mosaicism, where some cells in the developing embryo will be carrying the transgene and some will not.

In the first and second cases, the transgene will be inherited by offspring in a reliable, consistent and predictable way, for position and copy number. Segregation and random assortment of chromosomes at meiosis will not have any impact on the mode of inheritance of the transgene, because integration, regardless of copy number, has taken place at one site only. In the case of mosaicism, the transgene may be passed on to some offspring, but not all. However, F1 offspring that carry the transgene will do so in every cell of the body and transmission thereafter should be a predictable event.

Transgene expression levels in individual founder animals may be influenced by copy number. Integration site may also be influential, since regions flanking the transgene may affect expression levels (a “position effect"). All of the above are essential reasons to regard each founder animal as a separate line. They should be bred and maintained as such and evaluated as to which line is the most useful scientific tool.

As in pronuclear injection, each and every founder animal developing from an embryo generated with a lentiviral vector will be genetically different. In contrast to animals generated by pronuclear injection, lentiviral founder animals carry only one copy of the transgene at each integration site, and multiple integration sites on multiple chromosomes will occur in nearly all founders. This will have two consequences. Firstly, as for pronuclear-injected founders, each animal may express the transgene to a different degree, depending upon the integration site and copy number. Secondly, at each generation, copy number in offspring will change and will vary between offspring, as segregation and independent assortment of chromosomes takes place during gametogenesis. Therefore, in order to eliminate these variables, it is necessary to identify animals that express the transgene well and then go through several generations of breeding in order to establish a stable strain, each member of which carries one copy of the transgene that adequately expresses and is inherited in a stable manner.

11.6.2.1 Mating of F0 Animals

Since founder animals are all different, they should not be mated to each other, but outcrossed or backcrossed to your strain of choice. Founder animals, i.e., those that are born from the injected embryos, should first be analyzed by Southern blot, to determine how many copies of the transgene are carried by each founder. If it is possible to do so, also evaluate expression levels at this stage. More often, expression levels need to be evaluated in animals that do not need to be returned to the mating pool. Thus, mate each founder to one or two wild type mice. Genotype litters (F1 animals) by Southern blot (copy number still being an important factor), and evaluate expression levels.

11.6.2.2 Mating of F1 Animals

Future breeding stock should be selected from animals that have fewest copies of the transgene, but which still express well. Unlike transgenic animals produced by DNA pronuclear injection, multiple copies in lentiviral transgenics will have integrated on multiple chromosomes. Thus, at meiosis, genes may segregate out and, in doing so, give rise to a population of animals that may express the transgene at different levels. Thus, breeding the line to a population that carries only one copy of the transgene will stabilize the line. Again, mate to wild type mice, in order to try to develop a line that stably carries one copy only of the transgene. Animals arising from different founder animals should not be intercrossed. When mating transgenic mice to wild type animals, you can use either wild type littermates to breed with or you can backcross to an F1 animal or to an established inbred strain, such as C57BL/6. However, if you continue to do the former and use wild type littermates to maintain the colony, it is possible that the colony will begin to exhibit inbreeding depression, manifested as declining reproductive performance (see p 6 in Festing 1979). This can be avoided in the first instance by continually mating transgenic animals to hybrid or established inbred strains. In the event that the colony does become inbred by remaining "closed" for a period of time and breeding performance does decline, then a rescue can be effected by mating animals carrying the transgene to hybrid or inbred animals at that time.

Acknowledgements. Grateful thanks to Tom Gridley for the gift of CJ7 embryonic stem cells. Enormous thanks to Xin Yu and Diane D'Agostin for much work in generating the protocol for the preparation of chromosome spreads and in the generation and collection of data which formed the basis for statements made in Sect. 11.5.2. Also to Jade Wang for further development of the protocol for karyotyping. I am indebted, also, to Tom Fielder for critical reading of this manuscript.

References

Abbondanzo SJ, Gadi I, Stewart CL (1993) Derivation of embryonic stem cell lines. In: P. Wassarman, M. DePamphilis (eds) Methods in enzymology, vol 225. Academic Press, San Diego, pp 803–822

Auerbach W, Dumnore JH, Fairchild-Huntress V, Fang Q, Auerbach AB, Huszar D, Joyner AL (2000) Establishment and chimera analysis of 129/SvEv- and C57BL/6-derived mouse embryonic stem cell lines. Biotechniques 29:1024–1028, 1030, 1032

Bagis H, Odaman Mercan H, Dinnyes A (2004) Exposure to warmer postoperative temperatures reduces hypothermia caused by anaesthesia and significantly increases the implantation rate of transferred embryos in the mouse. Lab Anim 38:50–54

Champlin AK, Dorr DL, Gates AH (1973) Determining the stage of the estrous cycle in the mouse by the appearance of the vagina. Biol Reprod 8:491–494

Charreau B, Tesson L, Soulillou JP, Pourcel C, Anegon I (1996) Transgenesis in rats: technical aspects and models. Transgenic Res 5:223–234

Corbin TJ, McCabe JG (2002) Strain variation of immature female rats in response to various superovulatory hormone preparations and routes of administration. Contemp Top Lab Anim Sci 41:18–23

Doetschman T (1994) Gene transfer in embryonic stem cells. In: C. Pinkert (ed) Transgenic animal technology, a laboratory handbook. Academic Press, New York, pp 133–135

Festing FW (1979) Inbred strains in biomedical research. Macmillan, London

Howell RL, Donegan CL, Pinkert CA (2003) Mouse embryo yield and viability after euthanasia by CO_2 inhalation or cervical dislocation. Comp Med 53:510–513

Lemckert FA, Sedgwick JD, Korner H (1995) Gene targeting in C57BL/6 ES cells. Successful germline transmission using recipient BALB/c blastocysts developmentally matured in vitro. Nucleic Acids Res 25:917

Liu X, Wu H, Loring J, Hormuzdi S, Disteche CM, Bornstein P, Jaenisch R (1997) Trisomy eight in es cells is a common potential problem in gene targeting and interferes with germ line transmission. Dev Dyn 209:85–91

Lois C, Hong EJ, Pease SS, Brown EJ, Baltimore D (2002) Germline transmission and tissue specific expression of transgenes delivered by lentiviral vectors. Science 295:868–872

Longo L, Bygrave A, Grosveld FG, Pandolfi PP (1997) The chromosome make-up of mouse embryonic stem cells is predictive of somatic and germ cell chimaerism. Transgenic Res 6:321–328

Miyoshi K, Abeydeera LR, Okuda K, Niwa K (1995) Effects of osmolarity and amino acids in a chemically defined medium on development of rat one-cell embryos. J Reprod Fertil 103:27–32

Miyoshi K, Kono T, Niwa K (1997) Stage-dependent development of rat 1-cell embryos in a chemically defined medium after fertilisation in vivo and in vitro. Biol Reprod 56:180–185

Mullins LJ, Brooker G, Mullins JJ (2003) Transgenesis in the rat. In: A. Clarke (ed) Transgenesis techniques – principles and protocols. Methods in molecular biology, vol 180. Humana Press, Totowa, NJ, pp 255–270

Nagy A, Gertsenstein M, Vintersten K, Behringer R (2003) Surgical procedures. In: A. Nagy, M. Gertsenstein, K. Vintersten, R. Behringer (eds) Manipulating the mouse embryo – a laboratory manual. Cold Spring Harbor Laboratory Press, Cold Spring Harbor, NY, pp 258–260

of Laboratory Animal Resources, I. (ed) (1996) Guide for the care and use of laboratory animals. National Academy Press Washington DC pp 34–35

Robertson EJ (1987) Teratocarcinomas and embryonic stem cells. IRL Press Oxford chapter Embryo derived stem cells, pp 102–103

Robinson V, Morton DB, Anderson D, Carver JFA, Francis RJ, Hubrecht R, Jenkins E, Mathers K, Raymond R, Rosewell I, Wallace J, Wells DJ (2003) Sixth report of the BVAAWF/FRAME/RSPCA/UFAW joint working group on refinement and reduction in production of genetically modified mice. Lab Anim 37:S1.1–S1.51

Rough R (1993) The mouse, it's reproduction and development. Oxford University Press chapter Reproductive systems, pp 38–39

Schuster-Gossler K, Lee AW, Lerner CP, Parker HJ, Dyer VW, Scott VE, Gossler A, Conover JC (2001) Use of coisogenic host blastocysts for efficient establishment of germline chimeras with C57BL/6 ES cell lines. Biotechniques 31:1022–1024, 1026

Seong E, Saunders TL, Stewart CL, Burmeister M (2004) To knockout in 129 or in C57BL/6: that is the question. Trends Genet 20:2

Torres RM, Kuhn R (1997) Laboratory protocols for conditional gene targeting. Oxford University Press pp 57–72

Wurst W, Joyner AL (1995) Production of targeted embryonic stem cell clones. In: A. Joyner (ed) Gene targeting, a practical approach. Oxford University Press, Oxford, pp 33–61

Zhou Y, Galat V, Garton R, Taborn G, Niwa K, Iannacone P (2003) Two-phase chemically defined culture system for preimplantation rat embryos. Genesis 36:129–133

Appendix

Table 11.5. Embryos and cells, diameters and volumes

	Diameter	Volume	Injection volume
Zygote	73–85 microns	270 pl	n/a[a]
Pronucleus	15–22 microns	2 pl	2 pl
Blastocyst	84–96 microns	382 pl	10–15 ES cells
ES cell	10–15 microns	8 pl	10–30 fl
Perivitelline space	n/a	n/a	100 pl

[a] Not applicable

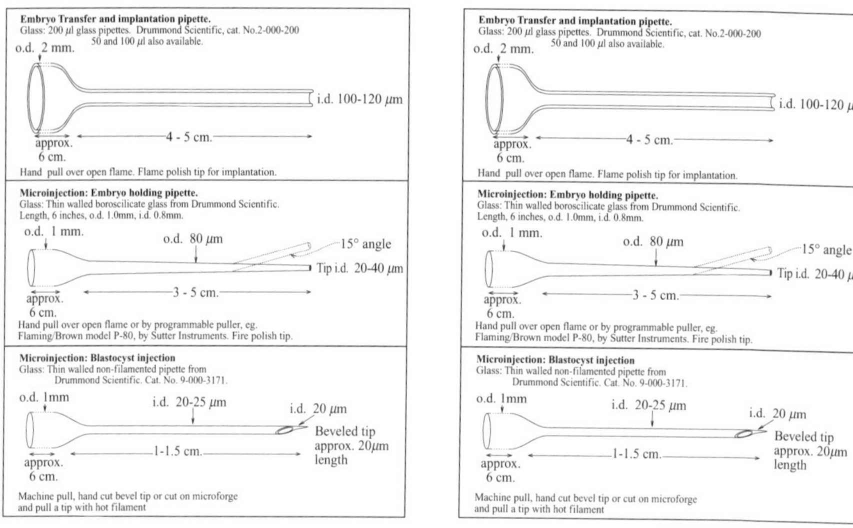

Fig. 11.7. A–F Microinjection pipettes. **A** Embryo transfer and implantation pipette. Glass: 200 μl glass pipettes (50 and 100 μl also available); Drummond Scientific, cat. no. 2-000-200; hand pull over open flame. Flame polish for implantation. **B** Microinjection: embryo-holding pipette. Glass: thin walled borosilicate glass. Drummond Scientific, cat. no. 9-000-3171. Length 15.2 cm, o.d. 1.0 mm, i.d. 0.8 mm. Hand pull over open flame or by programmable puller, e.g., Flaming/Brown model P-97 (Sutter Instruments). **C** Microinjection: blastocyst injection. Glass: thin walled non-filamented borosilicate. Drummond Scientific, cat. no. 9-000-3171. Machine pull, hand cut bevel tip or break on microforge and pull a tip with hot filament. **D** Microinjection: sub-zona (perivitelline space). Injection. Glass: as for blastocyst injection. Machine pull, hand cut bevel tip. **E** Microinjection: nuclear transfer by Piezo. Glass: as for blastocyst injection. **F** Microinjection: pronuclear injection Pipette. Glass: thin walled borosilicate glass, o.d. 1.0 mm, with internal filament. World Precision Instruments. cat. no. TW 100F-6. Machine pull

Subject Index

Printing: Krips bv, Meppel
Binding: Stürtz, Würzburg